Mines of
Silver Peak Range, Kawich Range and other Southern Nevada Districts

by S. H. Ball

This is a photographic reproduction of the 1907 U.S. Geological Survey Bulletin 308 entitled "A Geologic Reconnaissance in Southwestern Nevada and Eastern California."

Coverage includes
Esmeralda and Nye Counties in Nevada and Inyo County, California

Published by Miningbooks.com
For
Stanley Paher
Nevada Publications

PAHUTE MESA

Showing benches determined by resistant flows of basalt.

B.   DEATH VALLEY.

Showing sand dunes protected by desert shrubbery (mesquite and creosote bush).

CONTENTS.

ILLUSTRATIONS.

A GEOLOGIC RECONNAISSANCE IN SOUTHWESTERN NEVADA AND EASTERN CALIFORNIA.

By SYDNEY H. BALL.

INTRODUCTION.

FIELD WORK AND ACKNOWLEDGMENTS.

The field work on which this bulletin is based covered a period lasting from June to December, inclusive, 1905. The geologic work was under the general supervision of Mr. F. L. Ransome, and was carried on simultaneously with the mapping of the topographic base by Messrs. R. H. Chapman and B. D. Stewart.

The writer is fully conscious of the many inaccuracies of the geologic map herewith presented, but trusts that it may prove not only an aid to the prospector and miner, but a contribution to the geology of an intensely interesting region. Each day approximately 45 square miles was mapped in an area of by no means simple geology, and in many localities the geologic mapping of necessity preceded the topographic. No other region in the United States is so favorable to rapid mapping, since the exposures are unexcelled and the scant vegetation masks scarcely a geologic detail. The geology of the Paleozoic rocks is, however, difficult, partly on account of the lithologic similarity of many of the formations, but largely because these older rocks are in part hidden by mantles of Tertiary lavas and Recent gravels. In the text it has been the endeavor to distinguish clearly between observations made at close range and inferences drawn with the aid of the field glass. The mapping of the outer Goldfield hills and of the Pahute Mesa are particularly unsatisfactory.

To Mr. F. L. Ransome the writer desires to express his appreciation for general oversight and helpful criticism. Messrs. J. E. Spurr, F. B. Weeks, W. H. Emmons, and G. H. Garrey have been freely consulted on various points of Nevada geology. Messrs. R. H. Chapman and B. D. Stewart and their assistants, Messrs. T. C. Spaulding, E. A. Childers, and S. G. Benedict, aided the work by collecting in the field specimens of rocks from many points not visited by the writer. Mr. Spaulding also furnished some photographs and determined the species of the desert flora. To Messrs. W. F. Hil-

lebrand, George Steiger, and Waldemar T. Schaller the writer is indebted for mineralogical determinations, and to Messrs. E. O. Ulrich,
G. H. Girty, Edward M. Kindle, and F. H. Knowlton for paleon-
tologic data. The helpful interest and kindly hospitality of the Nevada pioneers is also gratefully acknowledged.

LOCATION AND AREA.

The area surveyed (see fig. 1) is situated in the south-central portion of the Great Basin and includes portions of Esmeralda and Nye

FIG. 1.—Index map showing location of area in southwestern Nevada and eastern
California.

counties, Nev., and Inyo County, Cal. The topographic map is on a
scale of 1:253,440 (approximately 4 miles to the inch), with contour
intervals of 100 feet, and covers the area lying between 36° 30′ and
38° north latitude, and 116° 00′ and 117° 30′ west longitude. This
area, embracing within its limits 8,550 square miles, is the equivalent of nine of the usual 30-minute quadrangles of the Geologic
Atlas of the United States.

LITERATURE.

The literature dealing with this portion of Nevada and California is scanty. In 1871 Mr. G. K. Gilbert visited the Reveille Range and later the Oasis Valley and crossed the Grapevine Range. In 1899 Mr. J. E. Spurr went from Lida to Stonewall Spring and thence across the Cactus, Kawich, and Reveille ranges. He also crossed Death Valley and the Funeral and Panamint ranges to the south of the area included in the present map. Since that time he has made extended studies in the neighboring districts of Tonopah and Silver Peak. In 1899–1900 Mr. H. W. Turner surveyed geologically the Silver Peak area, a thirty-minute quadrangle, which adjoins on the northwest the area mapped in this report. In 1900 Mr. M. R. Campbell made a reconnaissance to the south of the area here described. In 1905 Messrs. F. L. Ransome, G. H. Garrey, and W. H. Emmons made detailed studies of the geology and ore deposits of Goldfield and Bullfrog, and a preliminary account of the results of their work appears in Bulletin No. 303 of the United States Geological Survey. Since the discovery of the veins of Tonopah in 1900 the technical journals have contained numerous references to the mines in this portion of Nevada and California, but these have not been included in the following bibliography, since, with a few exceptions, they contain nothing of permanent value:

GILBERT, G. K. Report on the geology of portions of Nevada, Utah, California, and Arizona examined in the years 1871 and 1872. U. S. Geog. and Geol. Surveys West of the One Hundredth Meridian, vol. 3, Geology, 1875, pp. 21–187.
 Describes the Reveille and Amargosa (Grapevine) ranges and Bare Mountain, also the structure of the Basin Range.

HAGUE, ARNOLD. Geology of the Eureka district, Nevada, with an atlas. Mon. U. S. Geol. Survey, vol. 20, 1892.
 Describes the geology and ore deposits of Eureka, Nev. The Paleozoic section will be used as the type section in the present report.

SPURR, J. E. Descriptive geology of Nevada south of the fortieth parallel and adjacent portions of California. Bull. U. S. Geol. Survey No. 208, 1903.
 Describes the geology and contains a general geologic map, scale 15 miles to the inch.

TURNER, H. W. Silver Peak folio; unpublished.
 Describes general and economic geology of Silver Peak (Nev.) quadrangle.

SPURR, J. E. The succession and relation of lavas of the Great Basin region. Jour. Geol., vol. 8, 1900, pp. 621–646.

SPURR, J. E. Origin and structure of the Basin Ranges. Bull. Geol. Soc. America, vol. 12, 1901, pp. 217–270.

CAMPBELL, M. R. Reconnaissance of the borax deposits of Death Valley and Mohave Desert. Bull. U. S. Geol. Survey No. 200, 1902.
 Treats particularly of the borax deposits and of Tertiary lake beds.

SPURR, J. E. Geology of the Tonopah mining district, Nevada. Prof. Paper U. S. Geol. Survey No. 42, 1905.

SPURR, J. E. The ore deposits of the Silver Peak (Nevada) quadrangle. Prof.
    Paper U. S. Geol. Survey No. 55, 1906.
BALL, SYDNEY H. Notes on the ore deposits of southwestern Nevada and
    eastern California. Bull. U. S. Geol. Survey No. 285, 1906, pp. 53–73.
RANSOME, F. L. A preliminary account of Goldfield, Bullfrog, and other min-
    ing districts in southern Nevada; with notes on the Manhattan district
    by W. H. Emmons and G. H. Garrey. Bull. U. S. Geol. Survey No. 303,
    1906.

THE GREAT BASIN.

The Great Basin is an elevated region in Nevada and contiguous
portions of Oregon, California, and Utah, containing approximately
208,500 square miles. None of its streams flow to the ocean, and in
this it differs from all other provinces of the United States. The
region is arid, the precipitation being less than 20 inches a year.
The scenery, were it not for the grotesque form and bizarre coloring
of many of the mountains and hills, would be depressingly dreary. ·
.The mountain ranges of the region for the most part trend north
and south and are characteristically rugged and bare, although the
crests of some of the higher ranges are covered by a scant growth of
timber. The mountains are cut by deep canyons, and in a few of
these streams flow, only to sink in the desert gravels. Associated
with the ranges are low hill groups and mesas, many of which also
have a north and south elongation. Between the mountains and hills
are broad, gently sloping, inclosed valleys, which send branches into
and in places across the mountains. The lowest portion of most of
the valleys is occupied by either a lake or a playa, which during the
greater part of the year is a level waste of hard clay, but after heavy
rains is covered by a thin sheet of water.

TOPOGRAPHY.

The area under discussion, a typical portion of the Great Basin, is
one of mountain ranges and mesas with wide valleys of gentle slope
between. To the north of the broad Pahute Mesa lie mountain ranges
with a general north-south trend. Southeast of the mesa the mountains are small groups whose crest lines run in various directions.
The Grapevine and Panamint ranges, southwest of the mesa, extend
from northwest to southeast and are in consequence parallel to the
Sierra Nevada. The relief of the area is great; Kawich Peak is
9,500 feet above sea level and the part of Death Valley within the area
lies 280 feet below sea level. Few of the ranges, however, rise more
than 3,000 feet above the flat valley, although the mountain front
on either side of Death Valley reaches an elevation of 7,000 feet
within a distance of 12 miles. Between mountains with distinct
crests, such as the Kawich Range, and the small hillocks in the desert

valleys there is every gradation. The mountains are cut by gorges which in the more elevated ranges are deep canyons. The streams descend from the mountains to the flat valleys on alluvial fans and commonly even the stream channel disappears before the playa in the center of the valley is reached.

The valleys are inclosed basins which slope rather steeply next to the mountains, but decrease in grade rapidly, the central portion being a flat in which the eye can see no differences in relief. This level bottom is in most cases a playa. The borders of any given valley are, as a rule, approximately equal in elevation, although the desert gravels extend to greater elevations on the higher inclosing mountains. In such cases, while the slopes are of approximately equal descent, that from the higher mountains is much longer, and in consequence the playa is nearer the lower hills. Near the mountains the alluvial slopes are scored by numerous drainage lines, and hills protrude through their surface. The inclosed basins send bays and arms of detrital wash into the mountains and in places two opposed arms meet and form a strait of alluvial material. Every hill within the area furnishes each year material with which it is slowly being overridden by the constantly growing flat valleys. These valleys at one time may have been united to one another so that they formed a single drainage system, flowing to the south. Whether this was the case or not, it is evident that they are formed in three ways—by the union across a valley of opposed alluvial fans, as is exemplified by the small inclosed valleys of Grapevine Canyon; by the outflow of a lava barrier, such as that between Sarcobatus Flat and the inclosed valley to the north; or by orogenic movements, of which the Bullfrog Hills between Sarcobatus Flat and the Amargosa Desert are, in part at least, an example.

CLIMATE.

The area has an arid climate, with hot, dry summers and, except in the higher mountains, mild winters. The temperature in summer often rises above 100° F., the intensity of the heat, however, being mollified by the dryness of the atmosphere. Records kept at Hawthorne, Sodaville, and Palmetto, Nev., and quoted by Turner [a] show that the average annual precipitation to the north of the area varies from 3½ inches in the valleys to 15 inches on the highest mountains. These figures are probably both somewhat high for the area under consideration. The precipitation is largely concentrated in cloudbursts and 4 or 5 inches of rain may pour down in a few hours.

---

[a] Turner, H. W., Silver Peak folio; unpublished.

HYDROLOGY.

GENERAL STATEMENT.

The water resources of this area consist of streams, springs, tanks, wells, and snow. Data concerning the local water resources are given 'n the descriptions of the mountain ranges and valleys. While sufficient water is available or can be developed at a number of points for villages, mills, small ranches, and truck gardening, it is probable that the present supply of the area will never be greatly increased.

STREAMS.

Well-developed drainage systems exist in the hills and mountains, but nearly all the channels, except in times of cloud-bursts, are dry. Only in the higher mountains where heavy snows fall are perennial streams found, and the longest of these has a length of less than 4 miles. While the streams are small, a few, notably the Amargosa, furnish sufficient water for concentrating mills.

The so-called Amargosa River heads well up in Pahute Mesa, and the stream channel can be traced through the Amargosa Desert into Death Valley. Water comes to the surface only in that portion known as Oasis Valley, where numerous springs burst forth and give rise to small streams from 200 feet to one-half mile in length. Above Indian Camp the channel of the Amargosa occupies a canyon 500 feet deep, while between this point and Beatty the valley, from 200 yards to 1 mile wide, is bordered by low hills. The valley floor is here formed of fine clay, which, where not covered by drifting sand, supports a fair growth of salt grass. In places many acres are covered with an incrustation of white alkali. Between Beatty and Gold Center the valley contracts to a gorge. In the Amargosa Desert the channel is from 4 to 15 feet deep and 100 feet wide. In addition to the recent channel there are several older channel remnants. This shallow, sinuous channel, lined by heaps of angular bowlders and banks of sand, strongly resembles the deserted river beds of the Great Plains.

Four small streams are situated in Kawich Range. Breen Creek has its origin in a large number of springs rising from a marshy area on the Breen ranch. The stream in summer is from 1 to 3 feet wide and 2 to 4 inches deep. Early in the morning it flows to Silverbow, a distance of $3\frac{1}{2}$ miles, but by night the lower 2 miles are dry. The water at sunset is warm, while in the morning it is cold. The daily variation in size and temperature is characteristic of all the streams of this region and is due to the great heat of midday, which warms the water and induces evaporation. The large spring $1\frac{1}{4}$ miles below the Longstreet ranch yields a stream of water 1 foot wide and 3 or 4 inches deep, which sinks in the valley gravel one-half mile from its source. The underflow rises to the surface three

·times in the course of the tortuous Little Mill Creek Canyon, on the east side of the Kawich Range above Eden. The largest of these streams is 1½ feet wide and flows 1½ miles. A streamlet 200 yards long flows from the large spring commonly called " Georges Water," on the east side of the Kawich Range.

Salt Creek, 15 feet wide, rises in Death Valley, flows 2 miles, and then sinks. The water is heavily charged with sodium chloride and other salts. In the Salt Flat of Death Valley are many gently flowing small streams and pools of salt water. Small streams flow from the Staininger ranch and from Grapevine Springs, in the Amargosa Range, but each sinks within 1 mile of its head. Cottonwood Creek, in the Panamint Range, is 4 feet wide at its spring head, but sinks in gravel within 2 miles. Cottonwood and willow trees and grapevines line its banks, and water cress grows luxuriantly in the water. A short distance below the sink of this stream water again runs in the canyon for 100 yards. In the Stonewall Mountains water rises in the bed of a gulch one-fourth mile north of Stonewall Spring. The streamlet, 1 foot wide and 3 inches deep, sinks in the gravel 100 feet from its source.

SPRINGS.

The springs are of two kinds—hot or warm and cold. The hot springs appear to be vents of deeply circulating waters; the cold springs evidently come from shallow depths.

HOT OR WARM SPRINGS.

Alkali Spring is located 11 miles northwest of Goldfield. The waters originally rose at a number of small seeps, but recently the Combination Mines Company, of Goldfield, drove a tunnel into the gentle slope, concentrating the flow in a single channel. According to Mr. Edgar A. Collins, of this company, about 85,000 gallons of water per day flows from the spring and is pumped to the Combination mill at Goldfield. The water is clear, slightly alkaline in taste, and smells of hydrogen sulphide. An analysis by Abbot A. Hanks, kindly furnished by Mr. Collins, is as follows:

Analysis of water from Alkali Spring..

	Grains per wine gallon.
Silica, insoluble	2. 449
Iron oxides and alumina	. 314
Calcium carbonate	6. 764
Sodium chloride	6. 122
Sodium sulphate	43. 341
	58. 990

At the mouth of the 40-foot tunnel the temperature of the water is about 120° F., and at the breast it is at least 140° F. The stream flows from residual bowlders and soil of the later rhyolite, and the bowlders are badly decomposed and crumble readily in the hand.‘ One hundred yards north of the pumping station is a low dome of grayish-brown travertine, probably an abandoned vent of the spring.

Hicks Hot Springs, situated at the base of a low hill of silicified rhyolite on the east bank of Amargosa River about 7 miles above Beatty, are five in number. The hottest spring, which supplies the bath house, is said to have a temperature of 110° F. and to furnish 65,000 gallons per day. The springs in the vicinity which flow from the gravels of Oasis Valley are cool, and it is possible that the warm waters are genetically connected with the silicification of the rhyolite.

Water rises at the Staininger ranch, in the Amargosa Range, at several places in the canyon gravel, the total flow approximating 600,000 gallons a day. The water in December had a temperature of 75° F., while that of the air was but 49°. The surrounding hills are formed of Siebert lake beds and Pliocene-Pleistocene basalt flows. The Grapevine Springs flow from shallow depressions and valleys in the older alluvium and from the contact of that formation and the Pogonip limestone. Much of this water is tepid. Pliocene-Pleistocene basalt flows occur in this vicinity. At Ash Meadows, in the Amargosa Desert, 1 mile south of the area mapped, springs are associated with the older alluvium. The temperature of the larger springs is reported to be 76° F.; that of one of the smaller springs is 94° F.

These hot or warm springs are characterized by a rather large and constant flow of water. They are without much doubt the vents of deeply circulating waters, and their association in many places with recent volcanic rocks suggests that they may in part owe their heat to volcanism, if, indeed, they are not magmatic waters.

COLD SPRINGS.

Cold springs and seeps are situated either at the contact of alluvial deposits and the mountain bed rock or in the gravels of mountain canyons. In the latter case the surface waters appear to seep through the valley gravel and to rise where the bed rock forms a bar across the channel. The spring from which Cottonwood Creek flows may be cited as an example. Springs of this class are common in the large quartz-monzonite area of the Panamint Range. In the granite of Gold Mountain and the Cambrian rocks of the Montezuma district a water zone occurs at the contact of the solid rock and the residual and detrital soil. In other cases surface water has seeped into the more pervious strata, particularly the Siebert lake beds, the older alluvium, and altered varieties of rhyolite, and many springs are situated where these beds either outcrop or are covered by a veneer of detrital or residual material. Springs are located sporadically throughout the area surveyed, but seem to be more abundant around the bases of the higher mountains, especially of those which are covered with forests.

The precipitation both in winter and summer is greater on such mountains, and forested areas retard the run-off and permit the water to sink into the rocks. The flow is subject to seasonal changes, and during August and September many springs are dry or greatly diminished. Heavy rains also cause fluctuations. The water is cool, sometimes even cold, and is usually palatable, though a few of the springs are slightly saline, and some contain other salts.

From the direct dependence of the flow of the cold springs on seasonal rotation and on unusually heavy rains it is evident that these springs are in greater part, at least, the vents of shallow circulating waters.

TANKS.

Tanks are natural depressions in an impervious stratum, in which rain or snow water collects and is preserved the greater portion of the year. In the area studied the principal formations upon which tanks occur are the playa clay and the more compact layers of the basalt. The water in tanks decreases in volume and becomes poorer in quality as summer progresses; by August some are dry, and the water of others is scarcely drinkable.

WELLS.

Wells have been sunk at a number of places in the gravels of either the flats or the gulches. Wells sunk in gulches where there are signs of water have almost without exception proved successful; wells in the flats have been almost as uniformly successful. In the inclosed valleys the water table near the mountains, where the desert gravels are very permeable to water, seems to lie at considerable depths below the surface, but at the playas the two planes approach each other. In Death Valley water stands upon the surface, while in Sarcobatus Flat the water table is but 2 feet below the surface (fig. 2). In other flats the depth is considerably greater, but by analogy wells located near playas should encounter water at less depths than those on alluvial slopes. The supply seems sufficient for stock, but insufficient for extensive irrigation.

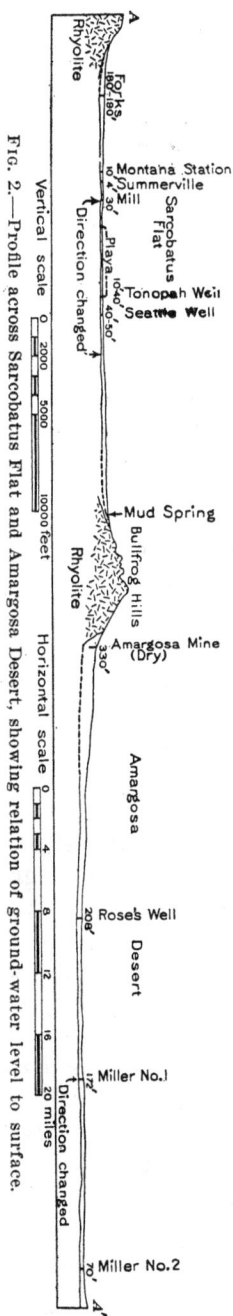

FIG. 2.—Profile across Sarcobatus Flat and Amargosa Desert, showing relation of ground-water level to surface.

The prospect of developing flowing wells in the area is slight. The Paleozoic rocks and the Tertiary volcanics are too much faulted to present extensive and continuous water-bearing strata. In the flat valleys beds of sand and gravel and layers of clay may be locally so superimposed upon one another as to furnish artesian conditions. Flowing wells, however, would be encountered only by the merest chance and would probably be weak.

In Turkestan and in the arid regions of southern California gently ascending tunnels have been driven into alluvial fans and water so obtained has been used for irrigation. The construction and maintenance of such tunnels are expensive, and in the area studied they would meet with success in few if any of the larger fans in which strong springs sink.

SNOW.

During the winter snow lies on many of the higher ranges and forms a valuable water supply to the prospector and traveler.

SIGNS OF WATER.

For the benefit of those unused to desert conditions it will perhaps be of value to describe the signs which indicate the presence of water in southwestern Nevada. Vegetation near springs or tanks is noticeably luxuriant, and in consequence patches of vivid green are worthy of investigation. The desert shrubbery attains an unusual size around springs, and a number of bushes and flowers not elsewhere seen are abundant. The willow, wild rose, elderberry, and gooseberry, the red honeysuckle and the poppy, and the " water bush," or tonopah of the Indians, are unknown except in the vicinity of water. The " water bush " is a shrub 2 to 3 feet high, with grayish-green leaves set on a white stalk thickly covered with spikes 1 inch long. Mesquite grows in Death Valley and the Amargosa Desert, and water may usually be obtained at slight depths near its roots. Where rye grass and the cane grow there is water at the surface or at moderate depths.

Rabbits, coyotes, and birds are particularly abundant near water. Mourning doves at sunset often flock around water holes. It is more than unfortunate that in the last two years the shotgun and trap have greatly diminished the numbers of this bird, one of the surest of water guides. Stock trails usually lead to water, and the point at which several trails converge is in most cases a spring. Since the animals go directly to water from the flats and on leaving the spring spread over the surrounding country as they feed, the trail becomes perceptibly plainer as it nears the spring. Many such trails, however, are winter trails, and in consequence the recency of tracks should be noted in searching for water.

Wagon roads and trails are constructed from one spring or tank to another and unwatered stretches of more than 40 miles are uncommon. The Indians, except when gathering piñon nuts, camp at water, and old Indian camping grounds, numerous fragments of stone implements, smoke-stained rocks, and carvings on rock may be valuable water indicators. Perhaps the best water signs are stone monuments built by the Indians and early white explorers either on high points on each side of the trail at water or close together on a commanding point above the water. These monuments are from 2 to 8 feet in height and resemble mineral monuments, except that they bear no location notices. From some of them flat stones set on edge point toward the water; at others a sharpened stick lying in a forked one is the indicator.

ANIMAL LIFE.

Animals are nowhere abundant in the area under consideration, and Death Valley appears to be without resident animals, with the exception of a few coyotes. The lack of food and the scarcity of water render the area as a whole unfit for large animals. Bands of wild horses roam the lowlands around the Cactus and Kawich ranges, however, and a few antelope feed in the valleys on either side of the Kawich Range and on Pahute Mesa. Mountain sheep are reported from the Lone Mountain country and old trails were noted on the higher peaks of the Stonewall Mountains. A few mountain sheep still live in the Panamint Range and in the south end of the Amargosa Mountains. Coyotes and rabbits are often seen, particularly in the vicinity of springs. Animals allied to gophers seem less dependent on known water holes.

Hawks, ravens, magpies, sparrows, and owls fly long distances from water. Mourning doves, on the other hand, flock around the springs and a few coveys of California quail live in the better watered portions of the Panamint and Amargosa ranges. Ducks in their annual migrations stop to rest and feed at the springs in the Oasis Valley and at Ash Meadows and the Furnace Creek ranch.

Various species of lizards, including the horned toad, are common in the valleys. Rattlesnakes, both the ordinary variety and the peculiar sidewinder, are sometimes seen, but are by no means as common as report would lead one to believe. The same is true of the scorpion and tarantula. Ants are common, and brilliantly colored butterflies are unexpectedly numerous. Mosquitoes and similar pests occur only near a few of the larger water holes.

VEGETATION.

Open forests of piñon (*Pinus monophylla*) and juniper (*Juniper utahensis*) occur above the dry timber line,[a] which approximates an elevation of 6,500 feet. The wood of both trees, particularly the

---

[a] Russell, I. C., Bull. Geol. Soc. America, vol. 14, 1904, pp. 556–557.

piñon, is good fuel. Only exceptionally are the trunks large enough for mining timbers other than lagging. The cones of the piñon incase delicious nuts, which are carefully gathered by the Indians for winter food. The vividly green mountain mahogany (*Cerocarpus ledifolius*) is associated with these trees on the highest peaks.

The most striking plant of the area is the yucca (*Yucca arborescens*), commonly called the " Joshua tree." It grows in open groves on the upper alluvial slopes and the lower lava mesas. Stiff spiny leaves are thickly set on the peculiarly branching stalk, which reaches a maximum height of 30 feet. Associated with the yucca are various species of cacti, including the prickly pear, the barrel cactus, and several branching forms. The sagebrush is widely distributed in the mountains, while the " creosote brush " (*Larrea tridentata*) and related shrubs occupy the lower alluvial slopes. In April and May the red geranium, the yellow dandelion, the purple horsemint, and the white primrose enliven the monotonous greenish gray of the desert shrubbery.

Grateful spots of green mark the springs. Here the gooseberry, wild rose, poppy, honeysuckle, wild rye grass, and cane thrive, while willow bushes and oak shrubs grow in thickets around some of the springs. At Poison Spring, in the Amargosa Range, the rocks are covered by the dainty maidenhair fern. In Cottonwood Canyon and at the ranches in Oasis Valley willow and cottonwood trees grow to a height of 40 feet.

Sufficient grass for scant forage grows on the upper alluvial slopes and in some valleys in the mountains. Burros will live almost anywhere in the area, but horses do poorly on the best of feed. The most nutritious grass grows in small bunches and is commonly called " sand grass." The sun's heat cures it in summer and in consequence a scant supply is available throughout the year. The grass around the springs, usually called " salt grass " (*Distichlis spirata*), is of little value to work horses. The white sage (*Eurotia lanata*) grows in the southern half of the area. Sheep are said to thrive on it and horses eat it with apparent relish.

CULTURE.

INDUSTRIES.

Prospecting and mining are practically the only industries, and nearly every person within the area surveyed is dependent directly or indirectly on these pursuits. The miners are segregated in the towns and camps; the prospectors range from one group of hills to another. At the Longstreet and Breen ranches in the Kawich Range, the Staininger and Indian ranches in the Armagosa Range, and at several of the springs in Oasis Valley small acreages are irrigated and hay, melons, potatoes, and other farm products are raised. The finer detrital material of the valleys is an excellent soil, but water for irrigation is lacking. A few cattle pasture in the Kawich Range, and while small herds might possibly live at other localities, the beef required by the mining camps will for the most part be shipped in by rail or driven across from California.

TOWNS.

The towns of this desert mining territory typify both the push and energy of the western pioneer and the unstable conditions of new gold fields. With the report of a rich discovery a town of 100 tents rises in a day, only to disappear almost as rapidly if the new find falls short of expectations. The population of all the towns fluctuates greatly.

Tonopah, with a population of about 8,000, lies 4 miles north of the area and is the county seat of Nye County, Nev. It has a number of prosperous mines and is the shipping point for some of the northern and northeastern mining camps of the area under discussion. Tonopah is the terminus of a railroad and has electric light, city water, telephones; newspapers, both daily and weekly, banks, churches, and schools.

Goldfield, with a population of about 9,000, has valuable gold mines and is the present terminus of the Tonopah and Goldfield Railroad. Like Tonopah, it has most of the essentials of city life, and the number of substantial buildings is surprising in consideration of the fact that the first tent was pitched here in January, 1904. Columbia, a rival town, adjoins Goldfield and has its own post-office. Goldfield and Columbia are supply points for the prospectors and miners in the central portion of the territory here described.

Rhyolite, Bullfrog, and Beatty are rival towns of the Bullfrog mining district. Rhyolite and Bullfrog, whose main streets meet, have a total population of approximately 2,500. Each has its own post-office, water system, and newspapers, and the two have common telegraph and telephone systems. The towns were founded in January, 1905, and already contain a number of substantial stone, wood, and adobe buildings. Beatty, 4 miles east of Rhyolite, is situated on Amargosa River and has a population of 700. It has about the same improvements as Bullfrog and Rhyolite. These three towns are the supply points for the southern portion of the area. Gold Center, a tent town of 40 inhabitants, is situated 2 miles below Beatty, on Amargosa River.

Silverbow, on the east side of the Kawich Range, has a post-office and a weekly newspaper. Kawich is a mining camp and post-office on the west side of the Kawich Range. The number of inhabitants fluctuates and has varied from 50 to 400. Lida is a mining camp of the seventies revived. At present it has a population of 200. Thorp, more commonly known as Montana Station, is a post-office and stage station. Travelers can find accommodations for man and beast at

Ramsey Well, Klondike Well, Cactus Spring, Tokop, Cuprite, Currie Well, Miller Well No. 1, and Ash Meadows.

RAILROADS AND STAGES.

The Tonopah and Goldfield Railroad is a broad-gage line, connecting with the Union Pacific Railroad at Hazen, Nev. The Carson and Colorado Railroad in Owens Valley is from 25 to 45 miles west of the western border of the area. The San Pedro, Los Angeles and Salt Lake Railroad, commonly called the Clark road, is from 40 to 75 miles east of the area. A railroad has been surveyed from Goldfield to Bullfrog and two branches are projected from the Clark road to the same point. The position of the railroads is shown on the sketch map (fig. 1). A daily stage runs from Goldfield and another from Las Vegas to the towns of the Bullfrog district. The trip from Goldfield takes about fourteen hours, from Las Vegas thirty-two hours. Daily or semidaily automobiles make the run from Goldfield to Bullfrog in four to six hours. Lida and Goldfield, Kawich and Tonopah, and Silverbow and Tonopah are connected by daily stages.

ROADS AND TRAILS.

Good roads and trails cross the area in all directions and few countries so sparsely populated are so accessible. The road gradient is low, although in the mountains short stretches are steep. Teaming is heavy in the sand surrounding the playas and in some of the stream gravels. Roads on playa clays are smooth and hard except after heavy rains, when wagons sink hub deep in the sticky clay. Roads and trails are shown on the geologic map (Pl. I). Few places are impassable to a pack train, and a wagon can go anywhere in the lower hills and mountains.

STRATIGRAPHIC GEOLOGY.

It is proposed in this section to describe briefly the formations of southwestern Nevada and eastern California (fig. 3) and to outline the general distribution of each. They are described in detail under each geographic subdivision in the section on descriptive geology. It will be convenient to take the formations up in order, from the oldest to the youngest, irrespective of their sedimentary or igneous origin. Owing to the lack of continuity and the complex structure of the Paleozoic rocks, their description is less accurate than that of the younger formations. The stratigraphic units of the Eureka [a] (Nev.) section are used as type formations for the Paleozoic rocks.

PRE-CAMBRIAN.

The presence of pre-Cambrian rocks was not definitely determined. The Prospect Mountain quartzite (Lower Cambrian) of the Specter

[a] Hague, Arnold, Mon. U. S. Geol. Survey, vol. 20, 1892, p. 13.

Range contains pebbles of quartzite, jasperoid, and vein or pegmatitic quartz derived from pre-Cambrian formations. The mica schist of the Bullfrog Hills is pre-Ordovician, and although considered of Cam-

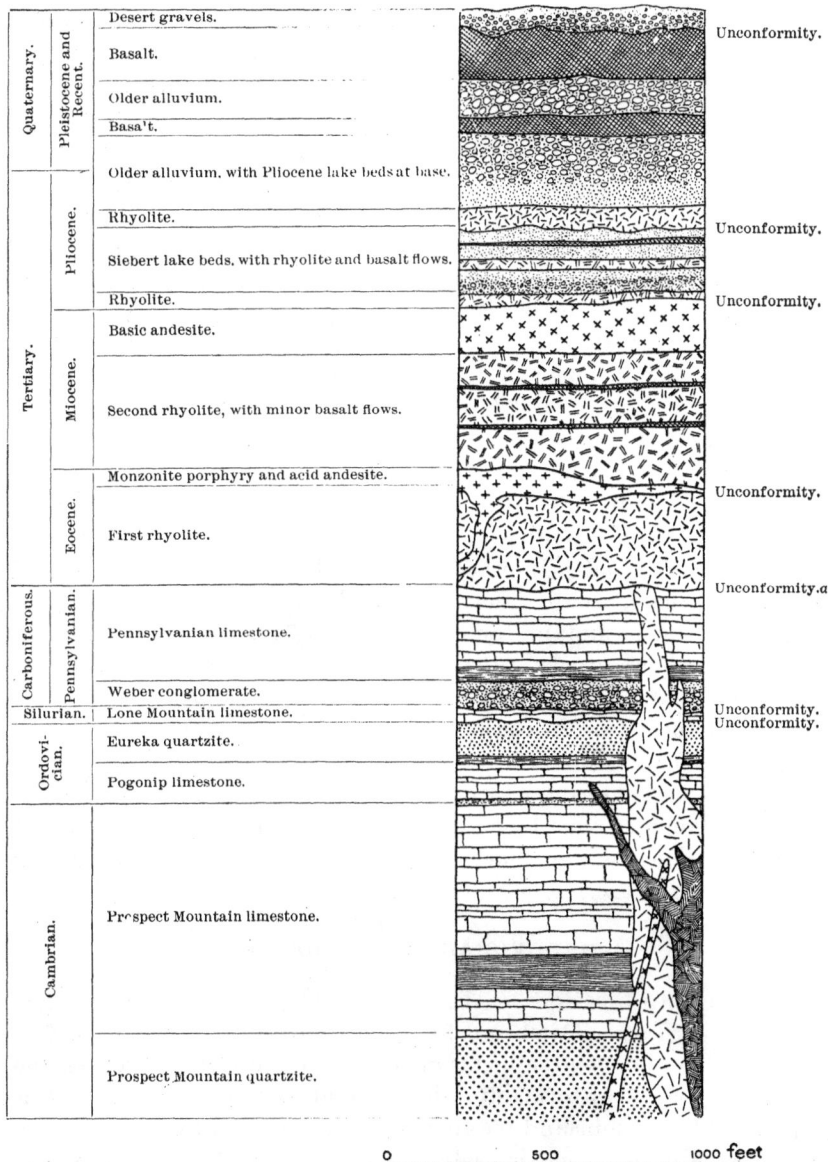

Quaternary.	Pleistocene and Recent.	Desert gravels.	Unconformity.
		Basalt.	
		Older alluvium.	
		Basa't.	
Tertiary.	Pliocene.	Older alluvium, with Pliocene lake beds at base.	
		Rhyolite.	Unconformity.
		Siebert lake beds, with rhyolite and basalt flows.	
		Rhyolite.	Unconformity.
	Miocene.	Basic andesite.	
		Second rhyolite, with minor basalt flows.	
	Eocene.	Monzonite porphyry and acid andesite.	Unconformity.
		First rhyolite.	
Carboniferous.	Pennsylvanian.		Unconformity.a
		Pennsylvanian limestone.	
		Weber conglomerate.	
Silurian.		Lone Mountain limestone.	Unconformity. Unconformity.
	Ordovician.	Eureka quartzite.	
		Pogonip limestone.	
	Cambrian.	Prospect Mountain limestone.	
		Prospect Mountain quartzite.	

0   500   1000 feet

FIG. 3.—Columnar section of rocks of southwestern Nevada and eastern California.

brian age, it may be pre-Cambrian. The schist at Trappmans Camp and the series in the south end of Amargosa and Panamint ranges are also probably Cambrian, although possibly pre-Cambrian.

a Rocks below this are cut by granite, which in turn is cut by monzonite porphyry, and this in turn by diorite porphyry.

CAMBRIAN.

In the Specter Range from 2,000 to 3,000 feet of quartzite and con-. glomeratic quartzite conformably underlie limestone containing Cambrian fossils. The rather impure quartzite, which grades into minor beds of slaty shale, is probably the equivalent of the Prospect Mountain quartzite [a] described by Hague, from the Eureka district. Less certainly to be correlated with this formation is the quartzite with intercalated schist and marble beds of the Amargosa and Panamint ranges, a rock series which may possibly be of pre-Cambrian age.

Conformably overlying the Prospect Mountain quartzite in the Specter Range is a limestone apparently from 5,000 to 6,000 feet thick, which is probably to be correlated with the Prospect Mountain limestone [a] of Hague. The limestone is dark gray, compact, fine grained, and crystalline. Black chert occurs at certain horizons. In the Silver Peak Range, Stonewall Mountain, the Goldfield, Southern Klondike, Lone Mountain, and Mount Jackson hills, and Slate and Gold Mountain ridges occur rocks that are in part at least of Lower Cambrian age. The predominant member is a dense dark-gray limestone of finely crystalline texture. As a rule it is massively bedded. In a number of places this limestone has been silicified to a banded gray and black jasperoid. Interbedded with the limestone are layers of green slaty shale, usually finely laminated. In the Silver Peak Range and the Lone Mountain foothills the shale reaches a thickness of 1,000 feet. These rocks, in part at least, are the equivalent of the Prospect Mountain limestone, although lithologically rather distinct from that formation in the type locality.

Of less certain Cambrian age are the mica schists of the Bullfrog Hills, Tolicha Peak, and Trappmans Camp. The Secret Canyon shale and the Hamburg limestone [a] and shale were not recognized, and the Cambrian limestones appear to pass upward without distinct lithologic break into the Pogonip limestone.

ORDOVICIAN AND SILURIAN.

The Pogonip limestone as described by Hague [b] is 2,700 feet thick and contains Upper Cambrian fossils at its base and Ordovician fossils in its middle and upper portions. At Eureka it overlies the Hamburg shale [a] conformably, while in southwestern Nevada it seems to succeed the Cambrian limestone without marked lithologic change. The Pogonip limestone is well exposed in the Belted, Amargosa,

---

[a] Present usage of the Survey does not sanction the double use of formation names, as Prospect Mountain quartzite and Prospect Mountain limestone, but as this reconnaissance paper is not the place to propose new names, they are retained as used by Hague.

[b] Hague, Arnold, Mon. U. S. Geol. Survey, vol. 20, pp. 48–54.

and Panamint ranges and in Bare Mountain, while transitional beds between it and the Eureka quartzite occur in the Kawich Range and a small area of what is probably the Pogonip limestone is situated in the Cactus Range. The limestone is dark gray, fine to medium grained, and dense. It is distinguished from the Cambrian limestones, already described, by a somewhat lighter color, a slightly coarser and less crystalline texture, and rather more massive bedding. Near the center of the section, which ranges in thickness from 2,000 to 4,000 feet, is about 100 feet of white or pinkish quartzite. The transitional rocks from the Pogonip to the overlying Eureka quartzite are an interbedded series of limestone, shale, and quartzite.

The Eureka (Ordovician) quartzite described by Hague as overlying conformably and without transitional beds the Pogonip limestone was recognized on stratigraphic grounds in the Amargosa and Kawich ranges, in the Bullfrog Hills, and on Bare Mountain, while in the Cactus Range and the hills between it and the Kawich Range quartzite areas isolated in Tertiary or Recent rocks are considered its equivalent. The Eureka quartzite in southwestern Nevada is a typically white or pink, fine to medium grained, pure metamorphosed quartzose sediment. Some beds, however, are conglomeratic quartzites, while others are argillaceous and grade into thin, interbedded sheets of dark-colored slaty shale. It varies from the quartzite of Hague chiefly in two characteristics—first, its greater thickness, reaching 1,200 to 1,500 feet in the Kawich Range, and, second, in being underlain by interbedded quartzites, shales, and limestones, transitional beds from the Pogonip limestone.

Hague [a] gave to the uppermost division of the Ordovician and Silurian rocks at Eureka the name " Lone Mountain limestone." This formation lies unconformably upon the Eureka quartzite and is 1,800 feet thick. It consists of black, gritty limestones at the base, which pass upward into light-gray, siliceous limestones. Its fossils include both Trenton and Niagara faunas. In the Amargosa and Kawich ranges and the Bullfrog Hills the Lone Mountain limestone is a gray, dense rock, of which about 400 feet is exposed.

DEVONIAN.

No Devonian formations were found in the course of the present work, and while such rocks may have been overlooked or may lie buried beneath Tertiary lavas it is tentatively believed that this portion of Nevada was a land mass during at least the greater part of Devonian time. The alternative hypothesis is that the Devonian beds were eroded away during post-Mississippian and pre-Pennsylvanian time, a period of important erosion in many portions of the West.

---

[a] Hague, Arnold, Mon. U. S. Geol. Survey, vol. 20, 1892, pp. 57–59.

CARBONIFEROUS.

Hague [a] describes the Weber conglomerate at Eureka as consisting of 2,000 feet of coarse and fine conglomerate containing angular and rounded fragments of chert, interbedded with reddish-yellow sandstones, and lying conformably between the limestones of the lower and the upper " Coal Measures." In the Belted Range the Pennsylvanian limestone is underlain by 300 to 500 feet of shale, which in turn overlies 800 to 1,000 feet of sandstone. The shales are thin bedded and of green, gray, or brown color. The sandstone is in part a rather pure quartzose rock and in part an arkose. Interbedded conglomerates contain well-rounded pebbles of limestone, quartzite, flint, and jasperoid, evidently derived from early Paleozoic rocks. Lithologically similar beds occur in the Cactus Range. These rocks are probably the equivalent of the Weber conglomerate at Eureka.

Pennsylvanian limestone also occurs in the Panamint and Reveille ranges and forms a small area on Shoshone Mountain. This limestone is dense, dark gray or black, and fine to medium grained. Specimens from many beds emit a fetid odor when hit with the hammer. Nodules of black flint are characteristic at many horizons. The limestone in the Belted Range, 2,500 feet thick, has near its center 55 feet of limestone conglomerate.

POST-JURASSIC AND PRE-TERTIARY IGNEOUS ROCKS.

The granite and associated rocks of the Belted and Panamint ranges, of Pahute Mesa, and of the Gold Mountain ridge contain fragments of gray, fine-grained monzonite and quartz-monzonite porphyry. These fragments, while clearly somewhat older than the rocks in which they occur, are massive and possibly represent an earlier solidification of the magma, from which the granitic rocks were afterwards differentiated. These rocks include types ranging on the one hand from alaskite through muscovite, muscovite-biotite, and biotite granites to a quartz monzonite approaching granodiorite, and on the other hand from a soda-rich granite to soda syenite. While the granitoid rocks in the areas shown on the map are of rather constant composition, transitions occur in a single area. In the Gold Mountain ridge, for example, of two ledges separated by less than 10 feet, one is a microcline-granite porphyry and the other a hornblende - bearing quartz - monzonite porphyry. Porphyritic types of the granitoid rocks occur, as well as aplites and pegmatites, later differentiations of the same magma.

These granitoid rocks are widely distributed over the region under consideration, although they cover larger areas in the western portion.

---

[a] Hague, Arnold, Mon. U. S. Geol. Survey, vol. 20, 1892, pp. 91–92.

Similar rocks are the predominant formation from longitude 117° 30′, the border of the region. westward to the Sierra Nevada.[a]

From the uniform amount of mashing which these rocks have suffered it is evident that all are approximately of the same age. They intrude the Paleozoic rocks where the two are in contact, and fragments of them are included in Tertiary lavas. The youngest of the Paleozoic rocks is of Pennsylvanian age and the earliest lava is probably Eocene; in consequence the granites are of post-Carboniferous and pre-Eocene age. It is shown in the section on geologic history (p. 39) that the close folding of the Paleozoic rocks occurred in post-Jurassic time and the granite intrusion is believed to have been a relatively late event in this period of deformation, an inference supported by the comparatively unmashed condition of the granites. Their post-Jurassic age is thus probable, a conclusion in accord with that of Spurr for the granites of Silver Peak[b] and Goldfield[c] and that of Louderback[d] for those of the Humbolt region. This would make these granitoid rocks .essentially contemporaneous with the granodiorite of the Sierra Nevada.

Dikes and sheets of a ferromagnesian-poor quartz-monzonite porphyry intrude Paleozoic sedimentary rocks and granite of the Silver Peak and Panamint ranges and of Slate Ridge. This rock near Lida is apparently cut by the diorite porphyry described in the next paragraph. It is presumably a second intrusion of the granitic magma.

Dikes of diorite porphyry and fewer intrusive masses of diorite occur in the Silver Peak, Panamint, Amargosa, and Cactus ranges, the Gold Mountain and Slate ridges, and the Bullfrog, Mount Jackson, and Lone Mountain hills. These rocks typically contain brown hornblende. They are younger than the Paleozoic rocks and the igneous rocks already described and nowhere were they observed cutting Tertiary lavas. Pebbles of diorite porphyry occur in the Siebert lake beds (Miocene) and the rock is probably of post-Jurassic and pre-Tertiary age. Its intrusion was the last event in the pre-Tertiary igneous activity.

TERTIARY.

The Tertiary rocks include a number of igneous rocks, lava flows and fewer intrusive masses, and sediments laid down in lakes. The accompanying table presents correlations of the Tertiary rocks of the various ranges

---

[a] Spurr, J. E., Bull. U. S. Geol. Survey No. 208, 1903, Pl. I.

[b] Spurr, J. E., Bimonthly Bull. Am. Inst. Min. Eng., 1905, No. 5, p. 955.

[c] Spurr, J. E., Bull. U. S.. Geol. Survey No. 260, 1905, p. 133.

[d] Louderback, G. D., Bull. Geol. Soc. America, vol. 15, 1904, p. 336.

The oldest of the Tertiary rocks is a rhyolite which occurs in Stonewall Mountain. This occupies a similar stratigraphic position with rhyolites in the Panamint Range, in the Randsburg district, and near Daggett in the Mohave Desert, which, according to Spurr,[a] lie beneath " lake beds which are probably, in part at least, Upper Eocene." This rhyolite in Stonewall Mountain is cut by dikes of quartz-monzonite porphyry and quartz syenite, which are probably contemporaneous with monzonite porphyry, quartz-monzonite porphyry, hornblende-biotite latite, and biotite andesite of other ranges. These rocks are in part intrusive masses and in part flows. Prior to the succeeding extrusion of rhyolite these rocks of siliceous andesitic and monzonitic composition were eroded, unconformities being observed in the Kawich, Amargosa, and Cactus ranges, in the Bullfrog Hills, and in Shoshone and Skull mountains. Then followed an extrusion of rhyolite, with minor siliceous latites and dacites, which was attended by insignificant basalt flows. This period of extrusion of rhyolitic lavas, the most important in this portion of the Great Basin, was not, however, strictly contemporaneous over the whole area; it is believed, for example, that the earlier rhyolite of Goldfield represents only the base of the rhyolite series of the Bullfrog Hills, and that the upper portions of the Bullfrog mass may have been extravasated while the andesites and dacites of Goldfield were solidifying. This rhyolite is everywhere separated from the Siebert lake beds by a marked erosional unconformity, and in the Kawich and possibly in the Cactus Range it is separated by an erosional unconformity from the succeeding andesites. The next younger igneous rocks, which are confined to the northern half of the area mapped, are basic andesites and dacites. These two rocks were found in contact only in the area covered by the Goldfield special map, and here Mr. F. L. Ransome[b] found the dacites intrusive into the andesites. The succeeding igneous rocks are rhyolites and siliceous latites and dacites which occur only in the Goldfield and Southern Klondike hills and the Silver Peak Range. In the latter two localities at least these rocks are interbedded with the Siebert lake beds without erosional unconformity.

Thick masses of sediments occur in the majority of the ranges of the area, and on lithologic and stratigraphic grounds are correlated with the Siebert lake beds of Miocene age at Tonopah described by Spurr.[c] These tuffaceous sandstones and conglomerates, largely composed of rhyolitic material, reach an observed maximum thickness (in the Amargosa Range) of 1,150 feet.

[a] Spurr, J. E., Jour. Geol., vol. 8, 1900, p. 633.

[b] Oral communication.

[c] Spurr, J. E., Prof. Paper U. S. Geol. Survey No. 42, 1905, pp. 51–55, 69–70.

Correlations of Tertiary formations in various ranges.

[Symbol × shows presence of formation.]

The ranges listed (column "Range."):

- Timber Mountain
- Mount Jackson
- Bare Mountain
- Monitor Hills
- Specter Range
- Yucca Mountain
- Revellie Range
- Lone Mountain foothills
- Panamint Range
- Hills between Cactus and Kawich ranges
- Death Valley
- Goldfield hills
- Belted Range
- Pahute Mesa
- Kawich Range
- Amargosa Range
- Bullfrog Hills
- Cactus Range
- Southern Klondike hills
- Stonewall Mountain
- Gold Mountain ridge
- Silver Peak Range
- Slate Ridge
- Tolicha Peak
- Shoshone and Skull mountains

Age	Formation	Presence of formation (×)
Eocene or earlier	Rhyolite	×
Eocene or earlier	Monzonite and biotite andesite	× × ×××× ×× ×?
	Erosional unconformity, Kawich, Amargosa, Bullfrog, Cactus, Shoshone.	
Miocene	Rhyolite, dacite, and latite	××× × ××××××× × ×× × ×
Miocene	Basalt, contemporaneous with rhyolite of preceding column.	×××
	Erosion between rhyolite and Siebert lake beds, and locally erosional unconformity between early Miocene rhyolite and Miocene andesite and dacite.	
Miocene	Andesite and dacite.	× × ×××× ×
Miocene	Rhyolite and biotite latite.	× × × ×
Miocene	Siebert lake beds.	× × ×× (c)×××××× (a) × ×
Miocene	Rhyolite and latite folded with Siebert lake beds.	× × ×××× ×
	Orogenic movement.	
Pliocene	Later rhyolite in places containing augite, not folded with Siebert lake beds.	× ×× × ×
	Mature land surface, probably here, at least, prebasaltic.	
Pliocene / Pleistocene in part	Later tuffs passing into older alluvium.	× × × ×× × ×?
Pleistocene in part	Basic andesites.	× ×
Pleistocene in part	Basalts in large part contemporaneous with upper part of older alluvium, but ranging from Miocene to Pleistocene.	××××××××××××××× × ×××××××

a Probably to north.

b Including shore formation.

c Shore formation only.

Before the deposition of the Siebert lake beds ceased rhyolites and siliceous latites and dacites outflowed. In some of the ranges these are folded with the lake beds; in others they lie upon the tilted and eroded surfaces of the sediments. During and immediately after the extrusion of these siliceous rocks tuffaceous sediments were deposited in several of the ranges in the northwestern part of the area, and these are in part probably equivalent to the Humboldt series, the deposits of the Shoshone Lake (late Pliocene) of King.[a] The older alluvium, so widely distributed, particularly in the inclosed valleys, represents ancient playa deposits and alluvial slopes of what was probably the Pleistocene residuum of this lake. The early stages of Shoshone Lake were followed immediately by basalt flows which cover wide areas in southwestern Nevada and eastern California. In several of the ranges these basalts were preceded by basic pyroxene andesites. While the basalt is thus largely of late Pliocene age, its extrusion began during the deposition of the Siebert lake beds, basalt flows occurring near the top of these sediments in the Amargosa Range and the Goldfield hills,[b] and probably in the Monitor Hills. Some of the basalt cones are remarkably fresh and are certainly of Pleistocene age, and it is by no means impossible that volcanism has not ceased in this portion of the Basin Range.

The succession of the lavas of the Great Basin has recently been treated at length by Spurr.[c] His succession for the petrographic province of the Great Basin and that of the writer for the portion of that province here treated are practically identical. Both assign the first rhyolite to the end of Cretaceous and to Eocene time. It is believed, however, from the length of time indicated by the complex history of extrusion and erosion intervening between this rhyolite and the deposition of the Siebert lake beds, that the monzonites and acidic andesites may well also be in part Eocene, while the second rhyolite is believed to be largely of early Miocene age. The andesite and dacite are also of comparatively early Miocene age, while the third rhyolite covers the middle and late Miocene and early Pliocene. The basalts range in age from late Miocene to Pleistocene, the major extrusions occurring in late Pliocene and Pleistocene time. As to the genetic relations of the magmas, the writer is wholly in accord with the views of Spurr in the article already cited. He believes that the Tertiary lavas of the Great Basin are the representatives of two complete cycles of the differentiation of a magma of medium composition into acidic and basic lavas and that probably the end of a still earlier cycle is also represented.

---

[a] King, Clarence, U. S. Geol. Explor. 40th Par., vol. 1, 1878, pp. 434, 456.
[b] Oral communication from Mr. F. L. Ransome.
[c] Spurr, J. E., Jour. Geol., vol. 8, 1900, pp. 621–646.

PLEISTOCENE AND RECENT.

TRAVERTINE.

At Alkali Spring, 11 miles northwest of Goldfield, is a low dome 200 feet long formed of platy masses of grayish-brown travertine. This was the vent of the spring probably in late Pleistocene time.

DESERT GRAVELS.

The desert gravels cover the surface of the broad valleys above the playas and fill valleys of erosion in the mountains. They extend over about 45 per cent of the area mapped. These detrital deposits are composed of bowlders, pebbles, and sand. Many of the angular or subangular bowlders near the mountains are 6 feet in diameter; near the playas the bowlders grade into pebbles, measuring one-half inch or less. Some bowlders of quartzite show small crescentic fractures, and some of fine-grained limestones show white crushed spots; these are due to the impact of one bowlder against another during rapid transportation such as is characteristic of cloudbursts. Near the mountains the gravels are but slightly stratified, and here cross-bedding and slight erosional unconformities are common. On the border of playas layers of small pebbles and sand are regularly interbedded with laminæ of playa clay. The material of these detrital deposits is undecomposed, and if consolidated would form a secondary rock of practically the same chemical composition as the rock from the débris of which it was derived. Calcium carbonate is a common cement particularly adjacent to limestone and basalt outcrops. In a number of the arroyos shelves along the sides and tiny scarps in the channels mark lime-cemented conglomerates. Gypsum and silica also occur as cements at a few places.

The thickness of the desert gravels is unknown. In some of the inclosed valleys lava flows and older alluvium are exposed here and there and probably underlie the valley at no great depth. In other valleys, however, the gravels probably reach a thickness of hundreds and possibly thousands of feet. The Amargosa mine at Bullfrog, which is situated about five-sixteenths of a mile from the rhyolite ridges, has been sunk 330 feet in detrital deposits. Wells have been put down in other valleys to depths of 75 to 275 feet, and, while the records are very imperfect, it is apparent that they have not passed through the desert gravels.

These gravels in large part make up alluvial fans and the contiguous alluvial slopes; in minor part they form cliff taluses. Where no playa exists there is a zone in the center of the valley in which the material is derived from the mountains on both sides, since cloudbursts, first from one range and then from the other, push the débris across the median drainage line.

The desert gravels on the surface are without question still accumulating and are in part of Recent age. Their deposition, however, is slow, and from their thickness it is probable that the lower portions of the formation are of Pleistocene age. The length of time required for their accumulation is indicated not only by the present infrequency of cloudbursts. Embedded in the gravels are pebbles deeply scored by wind erosion on the exposed side, while the under surface is untouched. Such pebbles must have been stationary for long periods. The alluvial fans, with the exception of a single channel, are superficially stained by the black " desert varnish," a process also requiring a considerable period.

PLAYA DEPOSITS.

Playa deposits composed of light-gray or white fine clay, in many places arenaceous, occupy the bottoms of nearly all the inclosed valleys. Grotesque concretions of calcium carbonate from 1 inch to 8 inches in length are locally embedded in the clay, while some strata are lime cemented. Salt Flat and other areas mapped as playa deposits in Death Valley differ from those just described in the fact that some water continually stands in depressions and that considerable amounts of saline material are being deposited. For convenience several areas in Oasis Valley have been mapped as playa deposits. These areas are occupied by clays, loams, and sands of complex origin, partly playa deposits, partly stream alluvium, and partly wind-deposited sand. The thickness of the playa deposits is unknown, although from their mode of origin it is believed that they are thinner than the desert gravels in the same basin.

During periods of unusual precipitation the playas are wholly or partially covered by water, and at such periods some fine detrital material is washed into them. More or less wind-blown sand becomes mixed with the clay, and on desiccation of the temporary lakes chemical deposits are formed. The playa deposits are of late Pleistocene and Recent age.

SAND DUNES.

Large sand dunes are confined to the southern portion of the area mapped, although smaller dunes border all the playas. Big Dune, in the Amargosa Desert, is 300 feet high, and lower ones, similarly perfect in contour, occur in Death Valley. These are said to retain their general position from year to year, their situation apparently being determined by wind eddies. Some of the smaller dunes are directly dependent on the desert shrubbery for their existence (Pl. II. *B*), while the position of those on higher ground is evidently determined by the protecting influence of surrounding hills.

DEFINITIONS OF IGNEOUS ROCKS.

The various igneous rocks which occur in the area under discussion are simply defined below for the benefit of the nontechnical reader.

Alaskite is a granular rock of white or light-gray color. It is composed almost entirely of quartz and orthoclase.[a] It is a granite lacking mica and hornblende.

Andesite is a porphyritic rock; that is, it has visible crystals (phenocrysts) embedded in a finer-grained, often glassy groundmass. The most conspicuous features are the phenocrysts of hornblende or biotite or augite and plagioclase.

Aplite is here applied to a finely granular rock which usually occurs in dikes cutting, for example, a granite. The aplite, which as a rule is more siliceous than the granite, in many instances has solidified from a portion of the granite mass that is still liquid at great depths.

Basalt is a black or dark-gray rock which is often porous through the presence of steam holes. Crystals of olivine, plagioclase, and pyroxene may or may not be visible in the dense or finely granular groundmass.

Dacite may be briefly defined as a quartz-bearing andesite. The more basic dacites closely resemble andesites, while siliceous dacites are closely connected by transition rocks to rhyolite and might be confused with rhyolite were it not for the presence of plagioclase crystals.

Diorite is a gray granular rock which is composed essentially of plagioclase and hornblende, while biotite or augite may also be present. Diorite and andesite are similar chemically, but differ texturally.

Diorite porphyry is a rock of the composition of diorite with abundant crystals of plagioclase, hornblende, and biotite. The groundmass is finely granular.

Granite is a granular rock of pink or gray color. It is composed of orthoclase, quartz, and either muscovite, biotite, or hornblende.

Granite porphyry is a rock of porphyritic habit having the same composition as granite. Quartz and orthoclase are the more common phenocrysts, and these are embedded in a finely granular groundmass.

Latite is a porphyritic rock with more or less glassy groundmass. Phenocrysts of orthoclase and plagioclase are present in equal amounts, while either hornblende, augite, biotite, or olivine is usually also present. Chemically, latite is the equivalent of monzonite.

Monzonite is a granular rock resembling both granite and diorite, but containing orthoclase and plagioclase in approximately equal

---

[a] Orthoclase (alkali feldspar) is unstriated; plagioclase (soda-lime feldspar) is striated and when light strikes it alternate tiny parallel bands reflect the light simultaneously.

amounts. If quartz is present, the rock is called a quartz monzonite. It is usually of a gray color.

Monzonite porphyry bears the same relation to monzonite as granite porphyry to granite.

Pegmatite is here applied to a coarsely crystalline rock which occurs as dikes in some other rock—granite, for example—with which it practically coincides in composition. Such rocks are probably formed by segregations from the original molten mass of the inclosing igneous rock.

Rhyolite is a porphyritic rock with glassy base. It has the same chemical composition as granite. The most abundant phenocrysts are orthoclase and quartz, although biotite, augite, and hornblende also occur.

Syenite is a granular rock usually of gray color. The component minerals are orthoclase and hornblende or mica. If quartz is present, the rock is called a quartz syenite, and with increase in this constituent quartz syenite passes into granite. Where the component minerals are rich in soda, the rock is a soda syenite.

GEOLOGIC HISTORY.

As has been stated, no pre-Cambrian formations have been found in the area studied, but the Prospect Mountain quartzite in the Specter Range contains pebbles which are in part derived from sedimentary rocks and in part from granite. It may be inferred, therefore, that pre-Cambrian rocks, both sedimentary and igneous, at one time covered areas in southwestern Nevada and eastern California.

In Cambrian and Ordovician time this portion of the Great Basin was part of a sea bottom upon which sandstones and then limestone and finally sandstones were laid down. The sinking of the sea bottom progressed step by step with the deposition of the sediments. King[a] has shown that the eastern border of the land surface from which the sediments were derived was, in the latitude of the fortieth parallel, near 117° 30′ west longitude, while the shore line farther south appears from Spurr's map[b] to have been at about 118° 15′. The early Paleozoic rocks of the area studied contain a greater proportion of sandy sediments, and the limestones are more impure than those described by Hague at Eureka, Nev., facts in harmony with the closer proximity of this area to the old continental shore line. The presence in the lower Paleozoic rocks of ripple marks, cross-bedding, intraformational conglomerates, and oolitic limestones indicates that the sea was for the greater portion of the time shallow, and locally, as evidenced by sun cracks, the deposits were even raised above sea

[a] King, Clarence, U. S. Geol. Explor. 40th Par., vol. 1, 1878, p. 247.
[b] Spurr, J. E., Bull. U. S. Geol. Survey No. 208, 1903, Pl. I.

level. The area under consideration was, indeed, so near the ancient
shore line, naturally a zone of greatest differential uplift, that islands
may have stood at times above the surface of the sea.

Hague[a] has found in the Eureka district a marked unconformity between the Eureka quartzite and the Lone Mountain limestone, and according to his description younger formations overlap the Eureka quartzite. Such an unconformity was not definitely recognized in the area surveyed, but it probably exists. The apparent absence of Devonian rocks has been already mentioned, and the alternative hypotheses are presented that the area was a land mass during the greater portion of Devonian time, or that Devonian rocks were deposited and later eroded. Spurr found Devonian rocks 18 miles south-southeast of Ash Meadows[b] and 5 miles east of Point of Rocks,[c] and his map shows large Devonian areas from 30 to 40 miles east of the area here considered. Whichever of these hypotheses may be true, it is evident that at some time between the deposition of the Lone Mountain limestone and the Weber conglomerate a land mass existed over a portion at least of southwestern Nevada and eastern California. Not only were the Cambrian, Ordovician, and Silurian rocks uplifted and eroded, but the limestones were cut by calcite veins and in part silicified to jasperoid, and the sandstones were cemented to quartzites prior to their inclusion as pebbles in the Weber conglomerate. Cementation is evident not alone from the presence of these pebbles, but also from the fact that the early Paleozoic quartzites are thoroughly indurated, while the Pennsylvanian sandstones are of more open texture. The earliest Carboniferous rocks exposed in the area studied are of Pennsylvanian age, and it is evident that during that period the area again formed the bottom of a sea, which, however, during the deposition of the Weber conglomerate could not have been far from a land mass, while a conglomerate in the Pennsylvanian limestone also indicates shallow-water conditions. After the deposition of this limestone the area was probably lifted above the sea and never sank beneath it again. The period of Paleozoic sedimentation thus concluded was unattended by volcanism, with the possible exception of that which produced the metamorphosed basic igneous rocks of the Amargosa Range.

King[d] believed that contemporaneously with the emergence of this portion of the Great Basin above the sea the old land mass to the west subsided and that upon it were deposited Triassic and Jurassic rocks. He further believed[e] that the ranges of western Nevada were formed at the same time as the Sierra Nevada, or at

[a] Hague, Arnold, Mon. U. S. Geol. Survey, vol. 20, 1892, p. 57.

[b] Spurr, J. E., Bull. U. S. Geol. Survey No. 208, 1903, pp. 197–198.

[c] Op. cit., p. 161.

[d] King, Clarence, U. S. Geol. Explor. 40th Par., vol. 1, 1878, p. 759.

[e] Op. cit., p. 747.

the close of the Jurassic. This seems very probable from the general parallelism of the Nevada ranges with the Sierra Nevada crest line, particularly since some of the ranges west of the area surveyed are composed of folded Triassic and Jurassic rocks. At this time the Paleozoic rocks were complexly folded, the larger proportion of the flexures striking west of north. The folding in some places is close and in others open, being particularly intense in the ranges nearer the Sierra, such as the Panamint and Amargosa ranges. It was accompanied by reverse faulting. These post-Jurassic structural lines have ever since their formation proved to be lines of weakness, of which advantage has been taken by later deformation.

After considerable folding had taken place, and probably before the end of the deformation, granitoid rocks were intruded into the Paleozoic strata. That the granite intrusion was a late event in the deformation is indicated by the massive character of that rock. The intrusion was accompanied by local buckling of strata, by contact metamorphism, and somewhat less directly by ore deposition. It was followed by the intrusion of a quartz-monzonite porphyry, and later of diorite porphyry and diorite.

A long period of erosion, extending at least to Eocene time, followed the post-Jurassic folding and consequent mountain building. This period was probably one of heavy rainfall and large rivers. No estimate can be formed of the amount of erosion, but it was sufficient to remove from the granite the covering beneath which it solidified. Notwithstanding the length of this erosion period no peneplain was formed; from this it is inferred that the area was subjected to successive uplifts while being eroded. Where exposed beneath the oldest Tertiary formation the Paleozoic rocks appear to have been eroded into mountain ranges, many of which have a north-south extension. Some of these mountains of Paleozoic rocks— for example, the Kawich Range—were fully as high and rugged as the present ranges.

The Eocene inaugurated the Tertiary period of volcanism and lake sedimentation processes, accompanied by important deformation, erosion, and ore deposition. The Tertiary igneous rocks, being largely in flows, in contradistinction to the intrusive post-Jurassic rocks, produced but little contact metamorphism. The permanency of the structural lines, initiated probably in early Cretaceous time, has already been mentioned. Along these lines the Tertiary deformation occurred, while many of the main lava extrusions burst from north-south vents along pre-Tertiary mountain ridges. The first lava to outflow was a rhyolite, which was followed by monzonites and acidic andesites, in part intrusive bodies. Next there was an important and probably long erosional interval, followed by a mighty extrusion of rhyolite, which equals or exceeds in bulk the basalt outflow of late Pliocene and early Pleistocene time. The rhyolite in turn was followed by andesites and dacites, and these by unimportant rhyolites, immediately preceding the deposition of the Siebert lake beds in the Pahute Lake. It was probably during this general period that the most important ore deposits in the Tertiary rocks were formed, presumably by waters heated by the still-hot magmas. Between the outflow of the rhyolite and the formation of the Pahute Lake erosion was probably continuous, partially accounting for the formation of the lake basin which in the main, however, originated through orogenic sinking that was possibly, in part, consequent on adjustments due to the extrusion of immense masses of lava, a hypothesis suggested by Spurr.[a] The Pahute Lake covered practically the whole area surveyed, although the presence of coarse conglomerates in the Kawich and Amargosa ranges and the Bullfrog Hills indicates that rugged islands rose above the surface of the lake. Wherever the former shores of the lake are now seen it is evident that the surface of the older rocks was uneven and that each island was surrounded by islets. The climate must have been moist and the presence of fossilized wood in the lake beds shows that trees flourished near its shores. While the lake was thus for the most part fresh, periods of aridity alternated with those of comparative humidity, and the lake or portions of it were partially desiccated, permitting the local precipitation of limestone, gypsum, and boron minerals. Volcanic flows and explosive eruptions of rhyolitic material occurred at various times during the existence of the lake. The Pahute Lake was destroyed in part by the increasing aridity of the climate and in part by deformation, which was accompanied and immediately followed by the extrusion of rhyolite. By this deformation the whole area was uplifted with attendant southward tilting, which accounts for the relatively low altitudes occupied by the Siebert lake beds in the southern portion of southwestern Nevada. Furthermore, the deformation, by differential uplift, blocked out the mountain ranges as they now appear and formed many of the inclosed valleys by broad folding or warping. Death Valley was at this time first outlined, though it was depressed later, probably in the late Pliocene or early Pleistocene time, by block faulting.

Extensive erosion followed and before the end of Pliocene time it had reduced the surface to mature and comparatively gentle topography. In restricted areas the Pliocene-Pleistocene basalt appears to have flowed upon a local peneplain. In late Pliocene time the climate was moist and a shallow lake probably covered a considerable area in the vicinity of Goldfield. The older alluvium, a formation widely distributed over the area, may be considered as deposits of the waning stages of this lake period in the inclosed valleys. Toward the

---

[a] Spurr, J. E., Prof. Paper U. S. Geol. Survey No. 42, 1905, pp. 52–53.

end of this period the basalt extrusion reached its climax. Uplift accompanied by normal faulting and folding followed the deposition of the older alluvium, and at this time erosion appears to have been very active.

In Recent time the erosion characteristic of arid lands has partially filled the inclosed valleys with bowlders, gravels, and sands—débris from the wasting mountains—and the process is still going on.

The ranges of this portion of the Great Basin were delineated by post-Jurassic folding and Cretaceous erosion. Later uplift, coupled with volcanic outbursts along these same structural lines, has added much to their height, and the effect of these processes has been emphasized by the deepening of the intermontane valleys by erosion. In Death Valley alone does faulting appear to have been the dominant process in blocking out the existing topographic forms. The unimportance of faulting in the topographic evolution elsewhere is well exhibited in the northeast quarter of the area surveyed. The Paleozoic rocks of the Kawich and the Belted and Reveille ranges are folded in a syncline with its trough in the Kawich and Reveille valleys. The presence of early Miocene rhyolites in the Cactus, Kawich, Reveille, and Belted ranges at approximately equal elevations further proves that, unless compensating faults are hidden in the intervening inclosed valleys, faults have been of little moment. The projection of minor transverse ridges from the main ranges into the desert valleys and the frequently observed gradation from small erosional buttes to groups of rounded hills, such as the Monitor Hills, and from these to such elevations as the Cactus Range, are also opposed to the view that the principal mountain masses have been bodily uplifted by great faults. The ascendency of erosion over faulting is further indicated by the broad and rounded detrital embayments which notch many of the ranges.

ECONOMIC GEOLOGY.

HISTORY OF MINING DEVELOPMENT.

Two waves of activity in prospecting and mining have passed over the portions of Nevada and California under consideration. The first started in the late sixties, reached its maximum height in the seventies, and died out before 1890. The hardy prospector and miner in these years confined his attention to the Paleozoic limestones and the post-Jurassic granites. Lida, Old Camp, and Montezuma were flourishing mining centers. Mining in the area studied was dormant in the nineties, but with the discovery of Tonopah's phenomenal veins in 1900 the desert region was attacked with new ardor. The Tertiary eruptives have been more especially the chosen field of the prospector, although the limestones and granites have been by no means neglected.

ECONOMIC CONDITIONS.

The various camps herein described are connected with supply points by wagon roads, most of them surprisingly good considering the fact that many are less than a year old. Some mines have a small but for the present sufficient water supply near at hand. Other

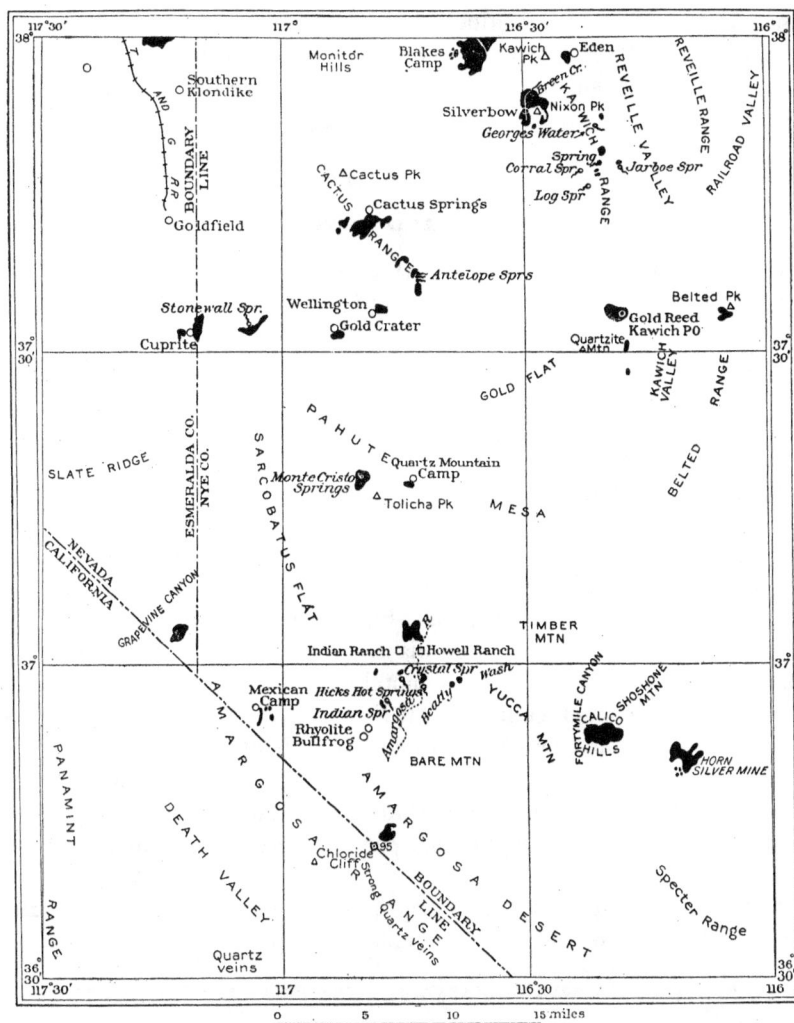

FIG. 4.—Economic map showing areas of altered Tertiary rocks.

camps are compelled to haul their water from springs or seeps 20 miles distant. Piñon and juniper are rarely of sufficient size for mine timbers, other than stulls, but piñon makes an excellent fuel. The power question will probably be solved at the other successful camps, as it has been at Goldfield, by the transmission of electricity generated on the swift streams of the Sierra Nevada. The heat of July and August renders surface work in those months almost impossible in the more southern camps.

GENERAL CHARACTER OF THE ORE DEPOSITS.

The ore deposits herein described are apparently the work of two distinct main periods of mineralization, one post-Jurassic and pre-Tertiary, the other Tertiary. The veins at Chloride Cliff may, indeed, be pre-Jurassic, but their resemblance to the post-Jurassic veins renders this questionable.

POST-JURASSIC AND PRE-TERTIARY DEPOSITS.

(A) DEPOSITS IN GRANITE.

1. Pegmatitic dikes: Dikes of pegmatite at Oak Spring are reported to carry gold and silver values. These, so far, have proved unimportant.

2. Reopened pegmatitic dikes: The quartz, itself of pegmatitic origin, has been crushed, and the veins are probably due to the same period of mineralization as that which formed the deposits described in the next paragraph. The deposits of Lime Point and some of those at Trappmans Camp are of this origin.

3. Massive quartz veins in fissures, joints, and irregular zones of brecciation: The quartz is not crustified. The chief sulphide is pyrite, although galena, chalcopyrite, and sphalerite occur locally. The deposits are usually gold bearing, but with the introduction of galena silver values may predominate. The ores were deposited by water, which may have been remotely connected with the granitic intrusion. The sharp contact with the comparatively unaltered wall rock indicates that these veins filled open fissures. Such appear to be the deposits at Old Camp and some of those at Trappmans Camp, Oak Spring, and Southern Klondike. The deposits of Goldbelt are in quartz monzonite. In the past the veins at Old Camp have been large producers.

4. Impregnation of pyrite along joints: This is an unimportant form of deposit seen only at Trappmans Camp.

(B) DEPOSITS IN PALEOZOIC SEDIMENTARY ROCKS, PREDOMINANTLY LIMESTONE.

1. Quartz veins and irregular masses occupying faults, joints, bedding planes, and brecciated zones, as a rule in the neighborhood of granitic intrusions: Rarely a little calcite is associated with the quartz. The original sulphides deposited in the quartz include chalcopyrite, galena, pyrite, and sphalerite, while telluride was probably originally present in some deposits. In the veins in limestone the predominant sulphides are chalcopyrite and galena and the

values largely silver. In quartzite and schist, in metamorphosed limestones composed of quartz, garnet, epidote, and calcite, and in limestone zones in which silification extended beyond the quartz vein pyrite is the predominant sulphide, with gold values alone or in excess of silver. It is probable that the relative abundance and variety of sulphides were determined by the wall rock, the more siliceous rock tending to precipitate from the waters auriferous pyrite and the calcareous rock precipitating argentiferous galena and chalcopyrite. The secondary ores of this subdivision include malachite, azurite, chrysocolla, native copper, brochantite, cerussite, emmonsite (or durdenite), native gold, and horn silver, while chloro-bromides of silver are reported. Chalcocite is probably of secondary origin. Two or more of these ores are associated with hematite and limonite or with a heavily stained jaspery quartz with conchoidal fracture. Gypsum, and at one place sulphur, are secondary gangues of less wide distribution. The secondary minerals replace the wall rock and in part fill open fissures. In many places cavities in the secondary ores are frosted by later quartz crystals. The Tokop veins belong to this subdivision and their similarity to veins in the granite of Old Camp and their reported extension into the granite beneath indicate that the quartz veins in the granite and limestone are contemporaneous in age and of like origin, a conclusion entirely in accord with the character of both the gangues and ores of types A 3 and B 1. The quartz veins in the granite (A 3) are not demonstrably of pegmatitic origin and yet their form, here lenslike, there sigmoid and independent of apparent fault-ing, is, in many instances, more closely allied to that of a pegmatite dike than to that of a water-deposited vein. In the Paleozoic rocks surrounding granite masses these quartz veins are abundant, a fact exemplified by the clustering of mining camps around the post-Jurassic intrusive masses. At a distance from the granite the veins decrease in number and in size, and finally the same ores replacing the limestone occur without quartz, as at Cuprite. It is probable that the quartz veins in granite and limestone were deposited by heated waters, in part at least the magmatic waters of the granite, but much more remotely connected with the magma than those which formed the pegmatite dikes. The re-semblance between the ores of these veins and those of the veins along contacts (B 3) is striking. The latter deposits may have been laid down by waters from portions of the granite deeply buried and still molten, or by waters from the white quartz-monzonite porphyry, an igneous rock probably genetically related to the granite.

The Lida, Bare Mountain, and Oriental Wash deposits, and some of those of Chloride Cliff, Montezuma, Oak Spring, and Southern

Klondike, are of this type. These deposits were the most important source of ore in the seventies and eighties.

2. Replacement deposits: In the Cuprite mining district and to a less extent in the General Thomas mine, in the Lone Mountain foothills, the sulphides appear to replace the limestone in irregular masses. The Cuprite deposits are at a considerable distance from granite, and in these quartz is lacking. The same ores are found here as in other deposits in the sedimentary rocks, and it is believed that the ore deposits of Cuprite were formed contemporaneously with those of the type B 1 by similar waters, which, however, had deposited their silica before journeying so far from the granite mass.

3. Veins along contacts: Deposits along contacts occur in the limestone along dikes and sheets of monzonite and diorite porphyry in the northwestern part of the area mapped, and along a mass of hornblende schist at Chloride Cliff. The porphyries are almost certainly pre-Tertiary and the mineral association is so similar to that in the deposits already described (B 1 and 2) that there can be but small doubt of at least a remote genetic relation in the origin of the two. The General Thomas and some of the Montezuma prospects are of this type.

TERTIARY DEPOSITS.

(C) GOLD DEPOSITS IN SILICIFIED MONZONITE PORPHYRY AND RELATED ROCKS.

Gold occurs at Kawich and Gold Crater, respectively, in silicified and kaolinized monzonite porphyry and biotite andesite. When the igneous rock was silicified pyrite was deposited in open fissures in it and probably partially replaced it. Disseminated pyrite occurs beyond the silicified zone. Copper sulphide is unimportant, and galena is apparently not present. Kaolinization is probably a later process. During the surface alteration of the pyrite free gold lodged in the cavities of the altered rock and at Gold Crater was deposited along joints in the surrounding iron-stained andesite. The quartz-monzonite porphyry of Shoshone and Skull mountains is similarly silicified. The ores of Kawich and Gold Crater were deposited by ascending hot waters carrying silica in solution, and while the waters were similar chemically, it by no means follows that the deposits are contemporaneous, since the country rock in the one case was probably of Eocene and in the other of Miocene age.

(D) DEPOSITS IN RHYOLITE.

1. Silver- and gold-bearing quartz veins in silicified and kaolinized rhyolite: At Silverbow, Eden, Cactus Spring, Wilsons Camp, Stonewall Mountain, and Wellington vuggy quartz veins, many of them with crustification well developed, fill fault fissures, joint planes, and cavities of brecciation and solution. These veins grade along their strike into zones of silicified rhyolite carrying gold and

silver values. The ore deposits of these camps, if not demonstrably of contemporaneous origin, were formed by physically similar waters. The original sulphides include stephanite, iron pyrites, and chalcopyrite. Pyrargyrite, also noted, may be secondary. From the primary ores free gold, horn silver, malachite, azurite, limonite, and hematite are derived. The relative abundance of silver sulphides or of pyrite determines whether the veins carry predominant silver or gold values. In the same camp structurally similar veins may carry either gold or silver and in various parts of the same vein the proportion of the two metals differs greatly. A little calcite is associated with quartz in some veins, and at Silverbow and Eden quartz bodies suggestive of the crystal form of calcite probably indicate its former presence. Irregular zones or areas surrounding these veins are either silicified or kaolinized, the former process having been contemporaneous with ore deposition, the latter, in certain cases at least, subsequent to it. The contact with the vein quartz is in some cases transitional, in others sharp. In all the camps these zones of altered rock are much more extensive than the ore deposits, and ore deposition unaccompanied by either silicification or kaolinization has apparently not occurred. The silicified rhyolite is white in color, flinty in texture, and without prominent phenocrysts, although careful inspection shows the presence of quartz phenocrysts, while weathered surfaces show feldspar casts. In some instances the slightly smoky quartz phenocrysts alone retain their identity. Some of the altered rhyolite is so dense that the Indians have used it in making arrowheads. During the deformation of the rocks after silicification the brittleness resulting from the density of this conchoidally fracturing rock was an important factor in the production of numerous cracks which furnished channels to surface oxidizing waters. Cavities of solution and the interstices of brecciated zones are lined with clear quartz crystals or filled with quartz masses. The alteration has taken place along joints and faults and in bands between joints. The silicified rhyolite, being much more resistant to weathering than the unaltered rock, stands up in rugged walls and knobs whose extent is coincident with that of the silicification. The peculiar topography resulting is well seen in the domical hills three-fourths of a mile east of north of Silverbow. Here there are rugged walls from 1 to 5 feet high, while cross walls show that silicification has also occurred along minor fractures between the main joints.

Under the microscope the silicified rhyolite shows a mosaic of quartz or chalcedony notably uneven in grain, while coarser mosaics or limonite-lined cavities represent the biotite and feldspar phenocrysts, although in some specimens biotite is replaced by muscovite and limonite. The silicified rhyolite of the Cactus Range, south of the

Goldfield-Cactus Spring road, presents features of considerable interest. The groundmass is composed in part of secondary interlocking plates of orthoclase micropoikilitically inclosing quartz blebs and in part of a secondary microgranitic mosaic of quartz and orthoclase varying greatly in grain throughout the rock. Alunite in grains is scattered through the groundmass. A few orthoclase phenocrysts are unaltered, but most of them are changed to an aggregate of alunite crystals and quartz granules. The rock appears to have suffered through recrystallization, with the introduction of sulphurous or a related acid and probably some silica. A thin section of the altered rhyolite from the Calico Hills, near Shoshone and Skull mountains, is somewhat different. The original rock was a glass containing numerous tabular orthoclase crystals and deeply embayed quartz grains. Chalcedonic quartz has replaced some of the feldspar phenocrysts, and with it, forming pseudomorphs after other feldspar phenocrysts and occurring scattered throughout the groundmass, are tiny irregular specks and granular aggregates of a mineral too small for satisfactory determination, probably scapolite. It is evidently contemporaneous with the chalcedonic quartz, and its even distribution throughout the rock shows that the solutions which deposited it must have saturated the whole rhyolite mass. The kaolinized rhyolite is a white, chalky, rather incoherent rock. The feldspar phenocrysts pass to a white unctuous substance and are finally completely removed, forming casts, and the biotite changes to a silvery micaceous mineral.

These quartz veins, the fillings of open fissures, and the attendant silicification of the country rock are evidently due to ascending heated waters which carried, in addition to metallic salts, notable quantities of silica, while locally they contained salts of calcium as well as fluorine, sulphurous acid and chlorine. The presence of the fluorine is indicated by the alteration of biotite to muscovite; sulphurous acid is a constituent of alunite and chlorine of scapolite. The waters in part, then, were of magmatic origin and contained gases typical of waning volcanism. These ore deposits, except those in the rhyolite of Stonewall Mountain and that to the north of Cuprite and some of those in the Southern Klondike hills, occur in the early Miocene rhyolite. But at Wellington and Wilsons Camp silicification appears to affect not alone this rhyolite, but the succeeding formation, the biotite andesite. This shows that at these places either the silicification was dependent on the magmatic waters derived from andesite or it occurred long after the extrusion of the rhyolite, probably through waters ascending from warm or hot portions of the rhyolite buried at considerable depths. The presence of silicified reefs in the Siebert lake beds in the Silver Peak Range and on Shoshone Mountain is a phenomenon probably of later origin.

Ore deposits of this type are now being extensively worked throughout southwestern Nevada and eastern California. As they have many analogies with those of Goldfield and Bullfrog, other important camps will without doubt be developed.

2. Gold ores in fault zones: At Blakes Camp free gold has been panned from crushed rhyolite along a fault. The iron pyrite was probably originally disseminated in the crushed zone.

3. Gold ores along contacts of Tertiary lavas and Paleozoic limestone: A quartz vein at Southern Klondike is situated along the contact of rhyolite with Cambrian limestone. At the Happy Hooligan mine free gold occurs in decomposed rock at the contact of basalt with Pogonip limestone.

AIDS TO PROSPECTING.

Ore deposits in the area under consideration appear to be confined to the Paleozoic rocks, the post-Jurassic granitoid rocks, and the older Tertiary lavas. Of the Paleozoic rocks the limestones have so far been and probably in the future will be the most favorable country rocks. As an exception to the above, however, it should be stated that the Cambrian quartzite-schist-limestone series in the southern Amargosa and Panamint ranges contains strong quartz veins. Of the post-Jurassic granitoid rocks the granites appear to be more favorable to ore deposits than the quartz monzonites. Of the Tertiary rocks the first and second rhyolites and the andesitic rocks immediately preceding and succeeding the second rhyolite have alone up to the present time proved productive. The rhyolites younger than the Siebert lake beds have, however, in two known instances been silicified and slightly mineralized.

The various formations recognized in southwestern Nevada and eastern California are represented by appropriate colors and patterns on the geologic map (Pl. I), from which the approximate distribution of the various rocks may be determined.

In prospecting in the Paleozoic rocks the limestones should be given preference. Mining experience in this region indicates that the most favorable portions of the limestone are those nearest igneous rocks, not only the larger intrusive masses, but the dikes as well. Such dikes are common in the Cambrian rocks from Lida to the General Thomas Camp. The abundance of ore deposits appears to depend almost directly on proximity to igneous rocks, and to this Cuprite alone seems an exception. Most of the ore deposits in the Paleozoic rocks are found in well-defined quartz veins, but some deposits are marked by heavy "gossans" (hydrated masses of spongy iron ores of red or brown color).

The map (Pl. I) shows the areas in which the older Tertiary lavas are known to occur, and, while many areas have doubtless been missed in the present rapid survey, it is believed that the larger ones have been mapped. The ore deposits within these older rhyolites and andesites, so far as is at present known, are contained in the silicified and kaolinized portions. These are shown in fig. 4, although it is certain that many others exist which were not encountered in the course of the present work. In searching for such areas of altered rock, the peculiar rugged topography with its walls (commonly called "dikes" by prospectors) and pinnacles (see p. 47) will be helpful in many instances. The compact, flinty texture of the silicified rocks, already described (see p. 47), is very characteristic. In the Kawich Range, Paleozoic quartzite has been mistaken by some for silicified rhyolite. The quartzite, however, can be distinguished by its more glassy appearance as well as by the presence in it of rounded sand grains or pebbles. In both the silicified rhyolite and the monzonitic and andesitic rocks veins occupy fault fissures, and prospect trenches and pits located on such faults often bring good results. Within the silicified rhyolite in particular quartz veins should be carefully assayed, as in this quartz the values in the rhyolite camps usually lie. Further, a yellow scaly coating similar to that from Goldfield, determined by Dr. W. F. Hillebrand [a] as a basic ferric-alkali sulphate, appears to be confined to joints in the immediate vicinity of ore deposits in rhyolite.

DESCRIPTIVE GEOLOGY.

In this section the different localities are described in the following order: The mountain ranges and valleys in the vicinity of Goldfield; those to the north of Pahute Mesa; the mesa itself; the mountain ranges to the south of Pahute Mesa; the Amargosa Mountain system; Death Valley; and the Panamint Range.

REGION ABOUT GOLDFIELD.

In the northwest quarter of the area shown on the map, in the vicinity of Goldfield, are the Lone Mountain foothills, the Silver Peak Range, the Mount Jackson, Goldfield, and Southern Klondike hills, and Stonewall Mountain. The hills are characterized by a north-south extension, the trend verging toward northeast in several instances. Geologically they are marked by important developments of the Cambrian rocks and the Siebert lake beds.

---

[a] Bull. U. S. Geol. Survey No. 260, 1905, p. 137.

TOPOGRAPHY AND GEOGRAPHY.

Lone Mountain is a prominent craggy peak 9,121 feet high, situated immediately north of the northwest corner of the area mapped. While this mountain and its foothills are cut off from the Silver Peak Range to the southwest by the detritus-filled pass 7 miles northwest of Montezuma, the two mountain masses are geologically and topographically closely related. For convenience, the long ridge in which General Thomas Camp is situated will also be described here.

The Lone Mountain foothills reach a maximum elevation of 7,600 feet. They have rather steep slopes and show a tendency to elongation parallel to the strike of the Paleozoic rocks. The eastern ridge has a serrated crest line, due to the alternation of steeply dipping resistant limestones and soft shales. The Lone Mountain foothills are bare of timber and without water, although both are present on the mountain itself.

GENERAL GEOLOGY.

Lone Mountain is a batholith of granite which intrudes Cambrian sedimentary rocks, the predominant rocks of the foothills. The formations are as follows, the oldest being named first: Cambrian sedimentary rocks, post-Jurassic granite, pre-Tertiary diorite and diorite porphyry and associated serpentine, and older alluvium.

SEDIMENTARY ROCKS.

Cambrian.—The predominant formation of these hills is a series of limestones and shales which appears to consist of a lower shale member, at least 1,000 feet thick, and an upper limestone, several thousand feet thick, which in turn is probably overlain by a second shale member.

The limestone is compact and rather fine grained. It is dark gray or bluish gray in color, although in places weathered surfaces are stained yellow or brown by limonite. It is for the most part probably dolomitic, since an analysis of a limestone from the westward extension of the area, in the Silver Peak quadrangle,[a] showed 19 per cent of magnesia. Layers of black flint, from a few inches to 2 feet thick, are interbedded with the limestones near General Thomas Camp. In the same vicinity the limestone has been altered to a white cherty quartz and all gradations exist between this and the unsilicified limestones. White calcite veins traverse the limestone in all directions, while quartz veins are less common. These veins are not folded, and are doubtless younger than the folding.

---

[a] Turner, H. W., Silver Peak folio ; unpublished.

The shale is fine grained and thin bedded. It is in most places olive green, although locally brilliant rose or purplish pink, while red is imparted to weathered exposures by iron stains. Arenaceous facies are rare. Rather commonly the shale breaks down into pencils which are produced by the intersection of two closely spaced joint systems and bedding planes. Veins in the shale are usually quartz.

No fossils were found in these beds, but on lithologic grounds they are probably Lower Cambrian. Turner [a] so considers the extension of these rocks in the adjoining Silver Peak quadrangle.

Older alluvium.—The road from Alkali Spring to Lone Mountain passes through a gap in the ridge south of the General Thomas mine. The saddle and the low hills on either side are composed of fragmental material lithologically like the Recent alluvium deposits, except that it is somewhat weathered. Erosion has cut it into a number of low hills. This deposit is the remnant of an old alluvial slope, and is tentatively correlated with the older alluvium of Pliocene-Pleistocene age in Death Valley.

IGNEOUS ROCKS.

Granite.—Lone Mountain is a batholith of granite which has pushed up the Cambrian rocks on its border, and into these apophyses extend. The batholith covers about 14 square miles at the northwest corner of the area mapped.

The granite is light gray in color and of either medium or coarse grain. The constituents, which in coarse-grained types reach a maximum diameter of one-half inch, are white feldspar, smoky quartz, and biotite. Phenocrystic feldspars 2 inches long and one-half inch wide are locally present. In some exposures the constituents possess a parallel arrangement, probably original. Epidote films are developed along some joint surfaces, while on others flattened, distorted cubes of limonite after pyrite occur. Muscovite films cover planes of movement.

An excellent sheeting, the joints being from 1 to 3 feet apart, cuts the granite and imparts to many portions of Lone Mountain a sedimentary aspect. The granite weathers into rounded slablike joint blocks stained by limonite, and in consequence the mountain mass is yellowish white in color. The mountain itself has a very rugged topography characterized by knife-edge divides and deep-cut, steep valleys. The foothills and inliers in the wash have strong tendencies to dome form.

Under the microscope the granite is seen to be uneven grained and rather poor in both biotite and quartz. With orthoclase and microcline is a little oligoclase. Titanite rhomboids are very abundant, and the other accessory minerals are magnetite, apatite, and

---

zircon. Quartz shows rather strong undulatory extinction. The rock is somewhat altered, biotite changing to chlorite, orthoclase and microcline to muscovite, and oligoclase to zoisite.

The granite is cut by aplite and grades into pegmatite. The aplite is a white, fine-grained rock composed of white feldspar and some quartz. It forms dikes from one-half inch to 2 inches wide, which weather in relief. The constituent minerals of the pegmatite reach a maximum diameter of 1 inch. Some veins of pegmatitic quartz are also present, as is graphic granite, with characters one-fourth inch in length.

At the contact the Cambrian rocks are rather highly metamorphosed. The impure limestone is altered to a quartz-epidote rock in which brown garnet occurs locally. The microscope shows that magnetite and pyrite are also present. The shale is metamorphosed to a silvery knotted schist. In thin section it is found to be a chlorite-epidote schist in which chlorite, quartz, and orthoclase form an interlocking mosaic. Epidote with a little zoisite is equally abundant, although less evenly distributed. Magnetite grains and a few fibrous scapolite aggregates are also present. Veinlets and lenses of coarser quartz, epidote, and zoisite appear to form the " knots."

The granite cuts Cambrian rocks and is itself cut by dikes of diorite porphyry. It is presumably one of the post-Jurassic granites.

Diorite and diorite porphyry.—A small intrusive mass of diorite cuts the granite and another the Cambrian rocks northwest of General Thomas Camp. The diorite appears to be a granitoid rock composed of feldspar and hornblende. The microscope shows that the feldspar is a rather basic labradorite in crystals, with some orthoclase. The hornblende is rudely idiomorphic. Some quartz and magnetite are present, the rock approaching a granodiorite. Epidote and chlorite are secondary to plagioclase and hornblende.

The granite of Lone Mountain is also cut by dikes of a greenish-gray diorite porphyry, which breaks into angular joint blocks with a slight tendency to spheroidal weathering. The dikes, which are from 2 to 20 feet wide, weather in troughs beneath the surface of the granite.

Under the microscope the groundmass appears as small plagioclase laths, between which are some orthoclase grains. The phenocrysts include hornblende, plagioclase (labradorite-andesine), and titanite. Many of the simple plagioclase crystals, which are more or less altered to sericite, calcite, and chlorite, are twinned according to the Carlsbad law. The brown hornblende phenocrysts are partially altered to calcite, chlorite, and iron ores. Apatite and magnetite are accessories.

Sheets and dikes of diorite porphyry, with abundant phenocrysts of striated feldspar and fewer of hornblende, occur in the vicinity of General Thomas Camp, their strike coinciding with that of the Cambrian rocks. The largest masses reach a thickness of 25 feet. Similar dikes cut Cambrian rocks north of the summit on the road from Alkali Spring to Silver Peak.

Pieces of greenish-black serpentine are common on the hills formed of the older alluvium, although the rock was not seen in place. Under the microscope the magnetite and hematite masses associated with the serpentine are seen to be lacking in certain areas, and the texture rather suggests that of an igneous rock. In the Silver Peak quadrangle Turner [a] found a small mass of serpentine in the Lone Mountain foothills. The original rock of each of these occurrences was probably a pyroxenite, genetically related to the diorite.

The diorite and diorite porphyry are probably of post-Jurassic and pre-Tertiary age.

STRUCTURE.

The Cambrian rocks are strongly and complexly folded. The axes of the main folds course east of north; those of the minor folds, east and west. The folding is particularly intense along the border of the area here discussed. Fan folds and isoclinals are common, and in many places these are arranged en échelon. Numerous minor faults and zones of brecciation occur in the Cambrian rocks, but faults of large throw were not noted. The injection of the Lone Mountain batholith has induced in the near-lying sediments a strike parallel to its boundary.

While it is probable that the Cambrian rocks were somewhat folded prior to the intrusion of the granite, the major folding can probably be referred to that intrusion. The position of the older alluvium, 300 feet above the Recent alluvium deposits, shows that the range was considerably uplifted in Pleistocene time.

ECONOMIC GEOLOGY.

The location of the General Thomas mine, the property of the Tonopah-Belcher Mining Company, is shown on the map. When visited by the writer (June, 1905) the mine was closed down. Several inclines follow the steeply dipping beds of limestone and shale, into which in the vicinity are injected sheets of diorite porphyry. Cerussite, malachite, azurite, and chrysocolla in limestone, heavily stained by limonite, lie on the dump. The gypsum associated with the ore was evidently formed simultaneously with these secondary minerals. Galena and pyrite in calcite were the only original sulphides seen, but some copper sulphide must also occur. Lakes [b] states that lenticular bodies of "sand" carbonate of lead, containing nodules of unaltered galena, lay along the contact of porphyry and lime-

---

[a] Turner, H. W., Silver Peak folio ; unpublished.
[b] Lakes, Arthur, Min. and Sci. Press, vol. 88, 1904, p. 246.

stone as well as in the lime-stone itself. The ore is said to carry high silver values, a carload shipment in January, 1904, having netted $4,300. At that time the shaft was 150 feet deep. The resemblance of this deposit to some of those at Montezuma may be noted. The surrounding hills are bare of timber, and water is scarce. Tonopah and Goldfield are the supply points of these prospects.

The hills near General Thomas Camp have already been located. An abandoned tunnel 150 feet long has been driven into the diorite mass 1 mile northwest of the General Thomas. The tunnel was not visited, but ore collected by Messrs. Chapman and Spaulding consists of galena, azurite, malachite, and limonite, which appear to fill cavities in the diorite. In the older alluvium the writer found limonite float heavily stained by malachite. This doubtless comes from the hills to the north or south of the road.

NORTHEASTERN EXTENSION OF THE SILVER PEAK RANGE.

TOPOGRAPHY AND GEOGRAPHY.

The mountains north of Lida and south of the valley in which the Klondike Well is situated form the northeast tip of the crescentic Silver Peak Range, which lies mainly to the west of the area surveyed. These mountains are cut off from the Lone Mountain foothills by the detritus-covered pass 7 miles northwest of Montezuma; they are separated from Mount Jackson by a similar gap, and a basalt mesa bridges the interval between them and the Goldfield hills. There are three main ridges, which course N. 30° E., parallel to the predominant strike of the Cambrian rocks. The crest lines average 7,500 feet in height, although Montezuma Peak reaches 8,426 feet. The ridges are, for the most part, rather smooth, although where the strike of the Cambrian rocks crosses them rugged squared hills and pinnacles are common. The hills formed of the Siebert lake beds are conical peaks and gentle domes with broad valleys between. The north-south valley southwest of Montezuma Peak follows approximately the crushed zone along a fault of considerable displacement. Lida Valley is an interesting example of stream capture. The traveler going up the valley from the flat north of Slate Ridge first encounters a narrow canyon. One-eighth of a mile east of Lida the gulch suddenly opens into a gently sloping valley 1 mile wide. The youth of the canyon and the maturity of the valley suggest stream capture, and Spurr, who followed the road westward, believes that the upper broader portion originally drained westward to Fish Lake Valley.

The mountains are covered with piñon and juniper above an altitude which varies from 6,500 feet at Lida to 6,800 feet on the east side of Montezuma Peak. A little grass grows on the upper alluvial slopes and the inter-ridge valleys. An abandoned shaft at Montezuma is filled with good water, and a tunnel near by furnishes a little stock water tainted with ferrous sulphate. Alkali Spring has already been described (p. 19). Springs flow from the mountain slopes near Lida, and the water table in the canyon below the village is in many places but 30 to 100 feet below the surface.

GENERAL GEOLOGY.

The formations exposed in these mountains, from the oldest to the youngest, include: Cambrian sedimentary rocks, post-Jurassic granite and gray quartz-monzonite porphyry, white quartz-monzonite porphyry, diorite porphyry, andesite, older rhyolite and dacite, Siebert lake beds, younger rhyolite, and basalt.

SEDIMENTARY ROCKS.

Cambrian.—The dominant rocks of these mountains are limestones, shales, and quartzites, in part at least of Cambrian age, and in the series no break in the process of sedimentation is apparent. West of Montezuma Peak these rocks consist of 1,000 feet of limestone with some shale beds near its top, overlain by 1,000 to 1,200 feet of shale with some interbedded limestone layers. A bed of white quartzite of fine to medium grain, 40 feet thick, occurs near the middle of the upper shale member. Neither bottom nor top of this series is exposed. Quartzite is more prominent in the hills southwest of Alkali Spring, where it is as abundant as shale.

The limestone is dark gray, compact, and fine grained. The weathered rock is medium gray in color, except in mineralized portions, where the reds, yellows, and browns of iron stains are characteristic. Much of it is siliceous and grades into a black, dense secondary jasperoid of conchoidal fracture. The limestone is massively bedded, except where interlaminated with shale, where it is usually in rather thin beds. Gray lenses and nodules of flint form one-third of the limestone mass 6 miles north of west of Montezuma Peak. By the addition of clay the limestone passes into calcareous shale and finally into shale.

The thin-bedded slaty shales are of fine, even grain and of gray or olive-green color. In many places muscovite plates are present upon bedding planes, and semischistose facies are developed by further formation of muscovite. Where joint planes are closely spaced the shale breaks into pencil-like fragments. Locally weathered surfaces are stained red or brown by iron compounds. With increase in size of grain and in content of quartz the shales pass into dense argillaceous quartzites of greenish-gray to black color and fine to medium grain. Cross-bedding is in places well developed.

The character of the rocks and the series of alternations of various rocks across the bedding planes indicate that the Cambrian rocks were laid down in comparatively shallow water in which conditions favoring mechanical sedimentation on the one hand and organic and perhaps chemical sedimentation on the other hand alternated.

Spurr[a] found fossils southwest of Lida which were determined by Mr. C. D. Walcott to be of Lower Cambrian age. The limestone 4 miles west-northwest of Montezuma Peak contains closely concentrically laminated balls averaging one-half inch in diameter. Mr. E. O. Ulrich states that these resemble *Girvanella*, a calcareous alga, and are probably of Cambrian age. On the evidence of these fossils and the resemblance of the series to that at Cuprite the rocks are believed to be, in the main at least, of Lower Cambrian age. Turner[b] maps Ordovician rocks at one place near the border of the Silver Peak quadrangle, but such rocks, if they exist, were not differentiated during the present reconnaissance.

Siebert lake beds.—Prior to the deposition of the Siebert lake beds the Cambrian rocks had a rugged topography. These beds, 17 miles north of Lida, are coarsely conglomeratic and inliers of limestone protrude through them and outliers of lake beds occur on the Cambrian rocks. The lake beds cover a considerable area on the western border of the region under consideration, a somewhat smaller area near Montezuma Peak, and a small area 5 miles south of Alkali Spring. The northern boundary of the largest area is approximately correct as mapped. The southern boundary is less accurately drawn and some Tertiary igneous rocks may lie between the Cambrian rocks and the lake beds.

These beds, which are composed largely·of rhyolitic material, consist of typically incoherent, unevenly granular, well-bedded sandstones. When unstained by iron, they are white in color, but brilliant reds and yellows are characteristic of many areas. Fine-grained beds alternate with conglomeratic layers containing pebbles of Cambrian rocks, quartz-monzonite porphyry, diorite porphyry, rhyolite, and igneous rocks of andesitic and basaltic affinities. The pebbles are well rounded or semiangular. Microscopic examination of several slides shows the presence of quartz and orthoclase crystals in a pyroclastic matrix.

Locally the Siebert lake beds have been indurated by silicification along joints and in irregular bodies. Quartz, chalcedony, and opal fill cavities. The silicified portions weather into wall-like masses along joints and into grotesque topographic forms where silicification has been less regular.

Rhyolitic and dacitic flows are interbedded throughout the sediments, although they increase toward the bottom of the formation and dacite appears to underlie it.

[a] Turner, H. W., Silver Peak folio; unpublished.
[b] Spurr, J. E., Bull. U. S. Geol. Survey No. 208, 1903, p. 186.

The lake beds at Montezuma Peak are about 1,100 feet thick. On the western border of the area mapped they are probably 200 or 300 feet thicker. The formation was for the most part deposited in an inland lake, although some portions were possibly laid down subaerially by explosive volcanic action.

Mr. R. H. Chapman found a piece of silicified wood at a point 4 miles west of Montezuma Peak. Mr. F. H. Knowlton, who kindly examined it, states that it is wood of a deciduous tree not older than the Tertiary. These sandstones are correlated with the Siebert lake beds of Tonopah, described by Spurr.[a]

Travertine.—A small area of travertine at Alkali Spring, presumably of Pleistocene age, has already been described (p. 35).

IGNEOUS ROCKS.

Post-Jurassic granite, granite porphyry, and quartz-monzonite porphyry.—The dikes and small irregular intrusive masses of white fine-grained alaskite and granite porphyry 2 miles southwest of Alkali Spring are usually injected parallel to the bedding of the upturned Cambrian rocks, but here and there cut across it. The porphyritic habit is better developed at the contact and next to the border phenocrysts are in flow orientation. The rock breaks down into angular joint blocks, many of which are stained by limonite. A single thin section under the microscope proves to be a granite porphyry with microgranitic groundmass of quartz and orthoclase. Coronæ of micropegmatitic quartz and orthoclase surround the abundant phenocrysts of dominant quartz and orthoclase and subordinate biotite and plagioclase (oligoclase). Some phenocrysts are composed of quartz and feldspar in a graphic-granite intergrowth. The phenocrysts all show undulatory extinction, and in consequence the rock has suffered some deformation. Biotite has been completely altered to muscovite shreds, rutile and an iron ore forming simultaneously.

The dikes on the east slope of this hill are composed of medium-grained gray biotite granite. On weathering the granite breaks down into spheroidal masses. It includes numerous small Cambrian fragments, while basic segregations of biotite and probably hornblende are present. This rock proves on microscopic examination to have a granular, hypidiomorphic texture, plagioclase individuals being of columnar habit. The orthoclase, much of which is microperthitic, is usually twinned according to the Carlsbad law. The feldspars are partially replaced by kaolin and sericite and the biotite by chlorite. Zircon and magnetite are accessories. Fine-grained pink aplitic dikes, from 1 inch to 4 inches wide, cut the granite and locally fault the Cambrian inclusions. These more acidic dikes are probably genetically related to the main granitic intrusion.

---

[a] Spurr, J. E., Prof. Paper U. S. Geol. Survey No. 42, 1905, pp. 51–55.

Dikes in the Cambrian rocks, 6 miles west of south of Alkali Spring, are composed of a dense white granite porphyry with rather numerous small phenocrysts. These dikes are too small to show on a map of the scale employed in this report. Under the microscope the groundmass appears as a mosaic of quartz and orthoclase which are in micropegmatitic intergrowth over small areas. A little plagioclase is also present. The phenocrysts include orthoclase, usually microperthitic, quartz deeply embayed, and plagioclase. In some specimens rutile rods are so grouped as to simulate somewhat rudely biotite crystals, and it may be that biotite phenocrysts present in the magma were absorbed prior to its solidification. Magnetite is an accessory.

The mass of quartz-monzonite porphyry 4 miles west of south of Alkali Spring and the dikes extending from it into the Cambrian rocks to the northeast are composed of a gray rock of porphyritic habit in which numerous medium-sized phenocrysts lie in a finely granular groundmass. The constituents in the order of their abundance are feldspar, both striated and unstriated, biotite, and hornblende. The rock weathers into rounded blocks. Inclusions of Cambrian rocks are common, as are basic segregations. Microscopic examination shows a microgranitic groundmass, composed of plagioclase laths, orthoclase tablets, and quartz grains. The phenocrysts consist of plagioclase; orthoclase, much of which shows excellent zonal growths; ragged biotite tablets, and hornblende columns. Magnetite and apatite are accessory minerals. The quartz-monzonite porphyry is cut by dikes of diorite porphyry and white quartz-monzonite porphyry. It is believed that this rock is a variant of the granitic magma intruded contemporaneously with the granites of the vicinity.

The Cambrian limestone in contact with the quartz-monzonite porphyry just described has been metamorphosed to a white, medium-grained marble in which epidote and other lime silicates are developed. The shales near the granitic rocks 2 miles southwest of Alkali Spring have been metamorphosed to slates. A dense, greenish-gray slate shows under the microscope as a mosaic of quartz and orthoclase in which are shreds of sericite and chlorite. Locally the sericite and chlorite form large skeleton growths containing numerous quartz and feldspar individuals. Another facies is a dense, black slate of conchoidal fracture. In thin section this is formed of large, rounded areas of orthoclase lying in a ramifying network of orthoclase grains intimately cut by biotite and muscovite shreds. Imperfect andalusite crystals are scattered throughout the rock, and some of these have the black centers characteristic of the variety called chiastolite.

These igneous rocks intrude Cambrian rocks and occur as pebbles in the rhyolite and Siebert lake beds. They are believed to be members of the post-Jurassic granite series.

White quartz-monzonite porphyry.—In the vicinity of Lida dikes and fewer sheets of white quartz-monzonite porphyry are common. The three areas shown on the map include both sheets and complex intrusive masses filled with inclusions of Cambrian rocks, while many other areas are too small to be mapped on the scale employed here. The dike which cuts the gray quartz-monzonite porphyry 4 miles west of south of Alkali Spring has already been mentioned.

This quartz-monzonite porphyry is a dense, white or locally greenish rock, with rather abundant medium-sized phenocrysts, consisting of somewhat altered whitish feldspars, some striated and others unstriated, silvery mica, and rarely a quartz crystal. The central portions of the dikes and intrusive masses are more coarsely crystalline than the borders, and in certain instances approach a granitoid texture.

The rocks break into sharp joint blocks, which in many places are stained by limonite. Limestone in the vicinity of this rock is here and there silicified and indurated, although the metamorphism is slight.

Microscopic examination proves the medium-grained microgranitic groundmass to consist of orthoclase grains, plagioclase laths, and a few quartz anhedra. The phenocrysts of orthoclase slightly predominate over those of plagioclase (oligoclase-andesine). Kaolin and sericite are the alteration products. Biotite phenocrysts altered to muscovite or chlorite, with an iron ore and rutile in rods or sagenitic webs, are constantly present, while quartz phenocrysts, much corroded, appear in some thin sections.

The white quartz-monzonite porphyry is younger than the granite which it cuts in the area 4 miles west of south of Alkali Spring. Pebbles of the porphyry are included in the Siebert lake beds. It is believed that it is genetically related to the granite and is a later intrusion of the same magma.

Diorite porphyry.—Dikes of quartz-diorite porphyry cut the gray quartz-monzonite porphyry 4 miles west of south of Alkali Spring, and many dikes of diorite porphyry, probably of contemporaneous age, intrude Cambrian rocks near Lida. The quartz-diorite porphyry is a dark, rather fine-grained holocrystalline rock. Under the microscope the presence of biotite is revealed. Ophitic texture is well developed, plagioclase laths (some andesine and more basic) lying in a mesostasis of hornblende, biotite, and a little quartz. The rock weathers into sharply jointed blocks of dark color. The diorite porphyry is a greenish-gray rock of porphyritic habit, both hornblende and striated feldspars occurring as phenocrysts. Epidote is developed in the rock and forms crystalline felts along joint planes. Spheroidal weathering is characteristic.

The contact of the white quartz-monzonite porphyry and the diorite porphyry is poorly exposed 1 mile northeast of Lida, and the latter rock is apparently younger than the former. Pebbles of the diorite porphyry occur in the Siebert lake beds—a fact which leads to the belief that these dikes were intruded in post-Jurassic and pre-Miocene time.

Andesite.—The andesite masses 4 miles north of Lida were seen by the writer only at a distance, but Turner,[a] who mapped their western extension, believes the andesite to be of " early Neocene " age, and it is probably older than the formation described in the following paragraphs.

Rhyolites and dacites.—Thin flows of rhyolite are interbedded with the lower portions of the Siebert lake beds of Montezuma Peak and of the area 4 miles north of it. The lake beds seem to lie on a siliceous effusive igneous rock which forms a narrow north-south band to the east of Montezuma Peak. Small areas of rhyolite are exposed by the erosion of the overlying basalt mesa $3\frac{1}{2}$ miles east and 4 and 5 miles southeast of Montezuma Peak. A small poorly exposed mass of rhyolite—possibly a dike, to judge from the disturbed condition of the surrounding limestone—lies 5 miles north of west of Montezuma Peak. Rhyolite is also exposed $3\frac{1}{2}$ miles north of Lida.

The rhyolite flows in the Siebert lake beds are dense gray or pink rocks, in which are sparse and small phenocrysts of quartz, orthoclase, and biotite. Black glasses are also present. Flow banding is highly developed, and spherulites form freely in cavities and impinge against one another in the more dense facies. The spherulites, some of which are simple and others compound, are of radial structure. The larger ones are 2 inches in diameter. Lithophysæ occur in the area 4 miles north of Montezuma Peak. Under the microscope the glassy groundmass appears somewhat devitrified. The flows here contain numerous pebbles of Cambrian jasperoid and quartz-monzonite porphyry.

The siliceous eruptive rock east of the Siebert lake beds of Montezuma Peak is a red semivitreous rock in which some small phenocrysts of striated feldspar and fewer of unstriated feldspar and quartz are sparsely distributed. The microscope shows numerous plagioclase microlites in the glassy groundmass, and orthoclase is so subordinate that the rock is a siliceous dacite.

The rhyolite of the other areas mentioned has abundant phenocrysts of quartz, orthoclase, and biotite in a light-colored groundmass which under the microscope appears as a glass somewhat devitrified, showing flow banding and spherulites. In one slide a few plagioclase phenocrysts are present with those already mentioned. Zircon and magnetite are accessories. This rhyolite grades into a white waxy glass without phenocrysts in the area 4 miles east of Montezuma Peak.

---

[a] Turner, H. W., Silver Peak folio; unpublished.

The siliceous igneous rocks are probably all portions of a single series that is older, for the most part, than the Siebert lake beds. They are probably contemporaneous with the Tonopah rhyolite-dacite described by Spurr [a] as presumably of Miocene age.

Later rhyolite.—Lying upon the eroded surface of the Cambrian rocks and protruding from the Recent alluvial deposits in the vicinity of Alkali Spring are small exposures of a lavender lithoidal igneous rock, with fairly numerous glassy feldspar phenocrysts of medium size. These rocks closely resemble the later rhyolite underlying the basalt flows at Goldfield and are tentatively correlated with it.

Basalt.—A basalt mesa lies between the spurs of the Silver Peak Range and the Goldfield hills. For convenience its description is given with that of the hill group (p. 75). A basalt dike 50 feet wide cuts the Cambrian limestone near its contact with the Siebert lake beds on the Montezuma-Goldfield road. The basalt is a dense black rock containing medium-sized glassy olivine grains and small striated feldspars and black augites. Under the microscope the rock is seen to be a holocrystalline olivine basalt. The plagioclase is partially altered to calcite and the olivine to serpentine. Basalt also covers a considerable area 1½ miles southeast of Lida. This is evidently a flow, the red or black rock being full of vesicles and much of it having a ropy surface. This rock is contemporaneous with the basalt of Mount Jackson and Goldfield, and the dike near Montezuma is probably of the same age.

Five miles northeast of Lida a flow of andesite or basalt lies upon the eroded surface of the Cambrian limestone. It has a gray lithoidal groundmass in which large striated feldspar phenocrysts are prominent. Lithologically the rock resembles certain members of the basalt flow south of Goldfield, but the mature topographic form of its surface may more nearly ally it to the older andesite already described.

STRUCTURE.

The Cambrian rocks are for the most part in rather open folds with north or northeast axes, which are crossed by minor folds. The amount of dip as a rule is not great, although at many points in the isolated hill southwest of Alkali Spring the rocks are on edge and the folding appears close. Locally small isoclinal folds are present, particularly near intrusions of granite, and the northwest strike of the Cambrian 5 miles north of Montezuma Peak is probably due to the intrusion of the quartz-monzonite porphyry. The gentle character of the folding of much of the Silver Peak Range is in marked contrast with that of the Amargosa Range and Bare Mountain to the south and southwest. The folding, in part at least, preceded the intrusion of the granite, since this rock is intruded along

---

[a] Spurr, J. E., Prof. Paper U. S. Geol. Survey No. 42, 1905, pp. 41–43.

vertical bedding planes southwest of Alkali Spring. Small faults, usually reverse, are common in the Cambrian rocks, while zones of brecciation in the limestone indicate that differential movement has occurred at many places. The Siebert lake beds are approximately horizontal, although flexures and normal faults occur. The normal north-south fault west of Montezuma Peak has a throw probably amounting to several thousand feet. The uplift of this range since Miocene time has been great, since the Siebert lake beds, once beneath a lake, have been elevated and eroded into mountains.

ECONOMIC GEOLOGY.

Two mining camps, Montezuma and Lida, are located in this spur of the Silver Peak Range.

MONTEZUMA DISTRICT.

The Montezuma mining district was organized in 1867, and soon afterwards several mines were opened in the country west of Montezuma Peak. In 1886 a mill was erected in the gulch to the northwest of the mountain. Active work ceased in 1887, but at present a number of prospectors are reopening old properties and locating new claims. The district product, estimated at $500,000, was freighted in wagons to Belmont, 65 miles away. At present only a few shallow prospect holes and the dumps of the old mines are accessible for examination. The mineralized area lies to the west of Montezuma Peak, in Cambrian rocks.

In the old mines the chief gangue is quartz, with rarely a little calcite and kaolinitic material. Vugs filled with quartz crystals occur in the quartz. The ores at the surface are cerussite, malachite, azurite, manganese dioxide, and limonite, and associated with these and replacing them in depth are galena, chalcocite, and pyrite. The values are largely in silver, the gold values being uniformly low. Chlorobromides of silver are reported, but were not seen.

A prospect hole exposes a quartz vein having white quartz-monzonite porphyry as the hanging wall and limestone as the foot wall. Chalcopyrite is the original sulphide, and malachite with less limonite, azurite, and a little native copper are the secondary ores. The native copper forms thin sheets in cracks and small nodules in the vein. Malachite partially replacing limestone was noted at a number of other places.

At Montezuma copper and lead sulphides with quartz fill open fissures in limestone. Some, probably all, of the deposition followed the pre-Tertiary dike injection. Later the veins were crushed, and surface waters altered the sulphides to carbonates, oxides, native metals, and probably haloid salts. This chemical breakdown of the sulphides continues to the present day, as is shown by the fact that the water flc ing from an abandoned tunnel southwest of the mill at Montezuma deposits considerable limonite and is so charged with ferrous sulphate that animals will scarcely drink it.

In several of the older mines and prospects water, both good and bad, was encountered at a depth of 30 feet. The mines will probably require pumps, but abundant fuel surrounds them. Goldfield, the railroad terminus, is 9 miles distant, the road being fairly good.

LIDA.

The Lida mining district was organized August 28, 1871, and in the succeeding decade some rich surface pockets of horn silver and silver-bearing galena were removed. The ore, probably picked, ran from $500 to $1,000 per ton. It is said that the values decreased at depths of 200 to 300 feet. In the latter part of 1904 and the early part of 1905, the attention of mining men having been turned to southwestern Nevada, old mines were reopened and new locations made. Unfortunately, at the time of the writer's visit (December, 1905), the principal prospects were closed through litigation, and in many cases pumps and ladders had been removed. The prospects are situated partly outside the area surveyed and partly on the east slope of Mount Macgruder to the south and southeast of the village.

The Florida-Goldfield Mining Company's shaft is near the mouth of a gulch which joins the Lida Valley about one-half mile below the village. When visited, the shaft, 150 feet deep, was filled with water within 80 feet of the surface. The ore on the dump includes Cambrian limestone rather heavily impregnated with iron pyrites and pyrite inclosed in veins of coarsely crystalline white calcite and in white quartz veinlets. In some specimens galena and light-brown zinc blende are associated with the quartz veinlets carrying pyrite. Chalcopyrite, an apparently still more uncommon sulphide, is in places superficially altered to malachite and azurite. The limestone cut by the quartz vein has been more or less silicified. An old open cut above this shaft shows a zone of brecciated limestone 4 feet in width healed by innumerable connecting quartz veinlets in which the ores mentioned occur. On the dump of the Thanksgiving mine quartz masses and veinlets cutting similar limestone were examined. The quartz contains much pyrite and less chalcopyrite.

The sulphides of the Lida deposits are in part fissure fillings and in part impregnations of the country rock, while the oxidized ores appear to be largely replacements of limestone. The oxidized zone is, for a desert country, very shallow.

In the early days of mining in this district considerable bunches of oxidized ore were hauled to Austin and Belmont. It is scarcely probable that all these pockets have been found, and with the improved transportation facilities such deposits should pay well.

Some of the ore mined from the newly located prospects is reported to run from $100 to $500 per ton in gold and silver. Such assay returns, however, in at least the majority of cases, were obtained from picked samples and are of no value in estimating the economic possibilities of a prospect. The water level, to judge from the development work so far carried on in the Lida district, is comparatively near the surface. Much of the ore already taken out is refractory and would require milling and smelting. In this respect the Lida district is at a disadvantage in comparison with the surrounding districts, but on the hills surrounding the mines there is a fair growth of piñon and juniper, which will furnish, for a time at least, satisfactory fuel and some mining timber. Water, sufficient for mining and domestic purposes, flows from springs in and above the village and can be obtained in the Lida Valley at slight depths. A daily stage runs from Lida to the railroad at Goldfield.

OTHER MINERALIZED AREAS.

Quartz veinlets are abundant in the Cambrian quartzite and shale and are locally present in the limestone, although calcite veins are more characteristic of the limestone. The quartz and calcite veins were evidently formed at several periods, since in many places one set faults another and in turn has itself been faulted. Strong quartz veins occur in limestone heavily stained by limonite and hematite 5 miles north of west of Montezuma Peak. Quartz veins also occur in the limestone area 6 miles north of Montezuma Peak. Thin sheets of malachite replace limestone along joint planes 3 miles north of Montezuma Peak.

LIME.

A kiln 2½ miles west of south of Montezuma Peak burns Cambrian limestone. The lime, which is sold in Goldfield, is reported to be of good quality.

LIDA-GOLDFIELD VALLEY.

The Lida-Goldfield Valley lies between Mount Jackson and the Goldfield hills on one side and the Silver Peak Range on the other. A low detrital divide west of Mount Jackson separates it from the valley north of Slate Ridge. The lowest part of the valley lies at an elevation of 5,100 feet, and from a distance appears to be occupied by two small playas. On the east side of the valley, north of Mount Jackson, some intensely dissected hills, 50 to 150 feet above the valley bottom, appear at a distance to be formed of playa deposits of the older alluvium. It is also possible that the apparent playas are in reality eroded exposures of the same formation.

MOUNT JACKSON AND THE HILLS TO THE NORTHEAST.

TOPOGRAPHY AND GEOGRAPHY.

Mount Jackson is a prominent butte which rises 800 feet above its alluvial base. Northeastward from it extends a ridge on which are superimposed buttes of the Mount Jackson type, capped by resistant lava flows. At its north end this ridge is divided by a valley into two parts. These hills are separated from the Goldfield hills to the north by basalt lava flows. For convenience the mesa-capped hill east of Cuprite and the isolated hills 10 miles east of north of Mount Jackson will be described here.

Water is unknown in these hills, and with the exception of scattered groves of the tree yucca they are bare. Grass is fairly abundant on the upper portions of the encircling alluvial slopes.

GENERAL GEOLOGY.

The formations exposed in these hills, named from the base upward, are Cambrian sedimentary rocks, diorite porphyry, Siebert lake beds, rhyolite, and basalt.

SEDIMENTARY ROCKS.

Cambrian.—An area in the southern ridge and another at the north end of the northern ridge are occupied by Cambrian sedimentary rocks, and the same rocks also in part form the isolated hills in the valley traversed by the Goldfield-Lida road.

A compact fine-grained black or dark-gray limestone, locally silicified to a black jasperoid of conchoidal fracture, is the dominant member of the Cambrian series. Both rocks are intricately cut by white calcite veinlets. Layers of thin-bedded green slaty shale, in places schistose, are interbedded with the limestone, as are smaller amounts of white medium-grained quartzite. A total thickness of 3,000 to 3,500 feet of Cambrian rock is exposed. These sediments form gentle hills elongated in the direction of strike, the shales being characterized by few, the limestone and quartzite by many exposures.

Mr. E. O. Ulrich determined the following fossils collected by the writer near the Goldfield-Midway-Bullfrog mining camp, 2 miles south of west of Cuprite: *Olenellus*, near *O. gilberti; Hyolithes*, closely related to or identical with *H. americanus;* undetermined orthoid shell.

He states: "So far as this imperfect material will permit of determining age, it points to a Lower Cambrian horizon." *Girvanella*-like forms similar to those from the Silver Peak Range were also found in these Cambrian rocks. The Lower Cambrian age of portions of the series is thus demonstrated, although other portions may be somewhat younger. The rocks are correlated with those of the north spur of the Silver Peak Range and of the Lone Mountain foothills.

Siebert lake beds.—White, finely bedded tuffaceous sandstones and clays, 200 to 300 feet thick, underlie the rhyolite of Mount Jackson and that of the isolated butte 10 miles east of north of it. Small areas of the same formation, not shown on the map, occur with the rhyolite northeast of Mount Jackson. Rhyolite pebbles occur in these sediments. At the more northerly butte they contain also thin rhyolite flows and lie unconformably upon the folded Cambrian rocks, which were evidently low hillocks in the lake bottom on which the sandstones were deposited. On lithologic and structural grounds these sandstones are correlated with the Siebert lake tuffs at Tonopah, described by Spurr.[a]

IGNEOUS ROCKS.

Diorite porphyry.—A dike of dark greenish-gray diorite porphyry cuts the Cambrian limestone one-fifth of a mile northwest of the Goldfield-Midway-Bullfrog mining camp already mentioned. The dike is 20 feet wide and courses N. 35° E. The dike rock, much altered, weathers into iron-stained spheroidal masses. The limestone at the contact is slightly indurated, approaching a hornfels. Macroscopically this dike rock resembles pre-Tertiary diorite prophyry of the Silver Peak Range.

Rhyolite.—Rhyolite flows cover a large portion of these hills. The rhyolite contains small and scarce phenocrysts of quartz, glassy feldspar, and biotite, which however, in some facies are lacking. The groundmass ranges from pumiceous to compact glasses of conchoidal fracture and varies in color from white through gray to black. Perlitic cracks traverse the glass in all directions and reddish-brown spherulites are common. The smaller spherulites show both radiate and concentric structure, while the larger, which reach a diameter of 8 inches, are traversed by fine wavy lines parallel to a diameter, although they break into sectors of a sphere. Some of the spherulites are simple, although as a rule two or more are linked together. Chalcedony is abundantly developed in joint cracks. Thin sections under the microscope show spherulites, cavities filled with gas, perlitic cracks which may extend through the phenocrysts, flow lines, and other phenomena typical of rhyolites.

The flow beds are from 2 inches to 50 feet thick, the upper surface being in some instances corrugated like the ropy surface of basalt flows. The maximum determined thickness of the rhyolite flows is 400 feet, although the total thickness may be somewhat greater. Small isoclinal folds were formed during the outflow of the lava, which must have been rather viscous and which on meeting an obstruction was crumpled. In one case a glassy facies folded with thin bands of semiglassy phenocryst-bearing rhyolite exhibits interesting

---

[a] Spurr, J. E., Prof. Paper U. S. Geol. Survey No. 42, 1905, pp. 51–55.

features. (See fig. 5.) The more crystalline rhyolite was sheared and fractured along the limbs of the fold, while the glassy facies thickened at the troughs and crests. The thinner flows weather somewhat like a horizontal sedimentary rock, while the thicker show the rounded topographic forms of granite.

Minor rhyolitic effusions started while the Siebert lake beds were being deposited, although the major outflow was later. The rhyolite is probably to be correlated with the younger rhyolite of the Amargosa Range and is of late Miocene or early Pliocene age.

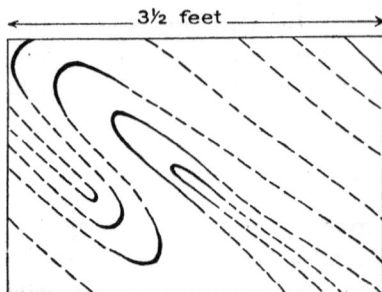

FIG. 5.—Flow folds in rhyolite (hill 10 miles east of north of Mount Jackson), showing thickening of more glassy rhyolite (white) on crests and troughs of folds and cross fractures in more lithoidal rhyolite on limbs of folds.

Basalt.—Slightly eroded benches of basalt flank Mount Jackson on the south and a small basalt area is located 5 miles to the east. The basalt is a dense black vesicular flow rock which microscopic examination proves to be an ordinary olivine basalt. As Spurr states,[a] erosion carved Mount Jackson approximately to its present form prior to the outflow of basalt. At the north end of the ridge east of Cuprite and in several small areas west of the Goldfield-Bullfrog road is basalt which is an extension of the younger basalt of the Goldfield hills. (See p. 75.) It is probable that small areas of younger rhyolite and late tuffs of these hills are here mapped with the basalt. The basalt northeast of Cuprite clearly overlies and is younger than the rhyolite, being probably of late Pliocene or early Pleistocene age.

STRUCTURE.

The Cambrian rocks are in rather open folds of northeast-southwest axes (fig. 6). The angle of dip rarely exceeds 30°, although

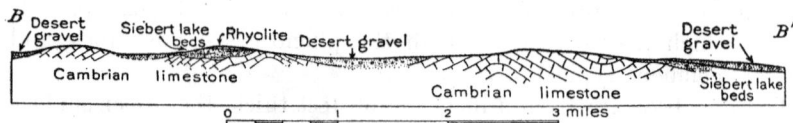

FIG. 6.—East-west section across Mount Jackson hills 1 mile south of Halfway Station.

4 miles west of Cuprite isoclinal folds occur. The highest hills of the two main ridges are situated on the axes of synclines. The rocks on the south face of the hills toward the valley north of Slate Ridge dip gently to the west. Faults, usually normal and across the strike

---

[a] Spurr, J. E., Bull. U. S. Geol. Survey No. 208. 1903, p. 183.

of the rocks, occur but do not seem to be of large throw. The Siebert lake beds and the later lavas are practically horizontal, their elevation being attended only by minor flexing and faulting.

ECONOMIC GEOLOGY.

The prospects of the Cuprite mining district, in which the first locations were made early in 1905, lie in a belt 1 mile wide which extends from a point 17 miles south of Goldfield westward to Mount Jackson. The mines are from 2 to 9 miles west of the stage and automobile road between Goldfield and Bullfrog.

The properties of the Goldfield-Midway-Bullfrog Mining Company and the Tri-Metallic Mining Company may be taken as types of the Cuprite mining district. The Copper Bell shaft of the first-named company, 85 feet deep, is located on the side of a gentle valley in Cambrian sedimentary rocks 2 miles south of west of Cuprite. The shaft is being sunk to encounter a vein which strikes N. 60° E. and dips from 80° to 85° NW. This vein has been traced by means of prospect holes at least 1,000 feet. At one prospect hole the vein is capped by a gossan of spongy limonite 9 feet thick. At this depth malachite and less azurite are associated with the limonite and at the bottom of the hole, 9 feet deeper, the vein is reported to have run 12 per cent copper, with 14 ounces of silver and $1.20 in gold per ton. The vein is from 2 to 5 feet wide and on its borders shows a gradual passage from totally unaltered limestone to pure limonite. The limonite, on the border of the vein at least, is a replacement of the limestone. The same company has an incline 65 feet deep situated on a shear zone which strikes N. 85° E. and dips 60° S., cutting practically horizontal limestone. A streak of ore $1\frac{1}{2}$ to $2\frac{1}{2}$ feet wide, traceable more or less continuously for 900 feet, occupies the shear zone. The ore consists of malachite, azurite, and limonite associated with heavily limonite-stained chalcedony of conchoidal fracture. The sulphides are chalcocite and less chalcopyrite and pyrite. The two latter are undoubtedly original. A number of assays of the ore are said to have averaged between 9 and 10 per cent of copper, while the silver values are variable, reaching as high as 400 ounces per ton. Evidently the silver is intimately associated with the copper minerals, the iron compounds giving no silver returns. The average ore runs about 1 ounce of silver per ton to 1 per cent of copper, while gold is present in traces only. Another prospect on this property is in limestone much stained by limonite. Malachite and less azurite, with an intensely limonite-stained chalcedonic quartz, occur in small kidneys throughout the limestone and in narrow seams along joint planes. Chalcopyrite is reported from the surface here, but is not abundant. In another prospect irregular lenses of the same ores and gangues lie in a narrow shear zone in the limestone. Small bunches of chalcopyrite are sporadically distributed in the dark chalcedony.

The property of the Tri-Metallic Mining Company lies one-half mile west of that last described. The Cambrian limestone, although gently flexed, is approximately horizontal. Intense staining by limonite is characteristic of limestone in the vicinity of the ore. A shipment of picked ore which netted $236 per ton (averaging 7 ounces gold, 230 ounces silver, and 19 per cent copper) was recently made from a surface pocket of ore. Chalcopyrite is the only original sulphide present. Chalcocite, malachite, and azurite with dark-brown chalcedony are secondary, and these are partially at least replacements of limestone. A few masses of native copper, presumably of secondary origin, are also reported, but they were not seen by the writer. Apparently chalcopyrite was originally distributed in the limestone in masses which, so far as known, do not exceed one-half inch in diameter. Later these scattered particles were concentrated, the secondary copper minerals apparently being replacements of limestone. In the main shaft bunches of chalcopyrite lie in the limestone in association with coarse calcite much stained by limonite. In some cases the chalcopyrite has been brecciated and the narrow interstices filled with iron-stained chalcedony. The ore in this shaft is said to have been in "kidneys," some of which were connected with others by narrow seams. From one of these masses 1,900 pounds of ore were removed. This lens of ore, 9 feet long, was 4 feet in diameter at the center and tapered to a point at each end.

Ore examined from a claim owned by E. Oldt showed similar characteristics, a little galena and white quartz of apparently equal age being associated with the sulphides. Transparent drusy quartz crystals contemporaneous with or later than the copper and lead carbonates and brown chalcedony are also characteristic. The values in this ore are said to be in silver and copper.

In the Cuprite district chalcopyrite and less pyrite, galena, calcite, and quartz appear to have been deposited as sporadic masses in the limestone as seams along joints and as lens-shaped bodies along shear zones. Later these ores were altered by aqueous solutions to chalcocite, carbonates, and oxides, the secondary ores in large part replacing limestone. At many points faulting accompanied or preceded the change. Contemporaneously a much iron-stained chalcedony was deposited in intimate association with the oxidized ores. Later, in some instances, drusy quartz was deposited upon the copper carbonates. The development is altogether too slight to determine what part of the secondary copper ores have been concentrated from sulphides originally lying in limestone now removed by erosion. It is probable, however, that a considerable proportion is of such origin and that when the oxidized ores are worked out the copper content will decrease. It may well be, however, that the water level at Cuprite is deep.

In the rhyolite north of the east end of the copper belt are some prospects. The values reported are gold, and these deposits are undoubtedly late Tertiary veins.

Water is hauled from Stonewall Spring, 17 miles distant. Wood, except the almost valueless yucca, is equally distant. The Cuprite mines, however, are only 18 to 25 miles from the railroad at Goldfield, and ore may be shipped from the district at a comparatively low cost.

Sulphur of good quality, said to occur in a vein in the rhyolite about three-fourths of a mile northeast of Cuprite, was examined.

VALLEY NORTH OF SLATE RIDGE.

The valley north of Slate Ridge is connected by two narrow arms of gravels with Sarcobatus Flat and by a single similar arm with Stonewall Flat. The valley has an east-west extension and is of moderate grade. Playas are situated in its eastern portion at elevations between 4,500 and 4,600 feet above sea level. Buttes of late Tertiary lavas are scattered over the lower portions of the valley. South of Mount Jackson a number of low white domes and dissected mesas of the older alluvium protrude from the Recent gravels, and the position of the more important of these is shown on the map. The exposures, where examined, consist of stratified clay beds lying horizontally. The area immediately south of Mount Jackson is 300 feet above the lowest deposits in the valley, and this may be taken as the minimum thickness of the formation, which is evidently a playa deposit of the older alluvium. The Recent gravels are probably thin throughout the valley. Some water lies most of the year in a tank at Forks Station.

GOLDFIELD HILLS.[a]

TOPOGRAPHY AND GEOGRAPHY.

The Goldfield hills are separated from the northeastern spur of the Silver Peak Range and the Mount Jackson hills by low mesas of basalt and from the Southern Klondike hills and the Cactus Range by low ridges of desert gravels. Most of the low hills and ridges around Goldfield have a north-south trend, but some run east and west. The highest summits are less than 7,000 feet above sea level. Except for the usual desert shrubbery the Goldfield hills are bare. Several springs occur in the eastern portion of the hills, and a spring

---

[a] The ore deposits and geology of the Goldfield special area have already been described by Mr. F. L. Ransome (Bull. U. S. Geol. Survey No. 303). The map made by Mr. Ransome and his assistants, Messrs. G. H. Garrey and Wm. H. Emmons, has been generalized and incorporated in Pl. I, and the writer is indebted to these gentlemen for many suggestions as to the stratigraphy in these hills. The description here given applies mainly to the Goldfield hills beyond the area shown on the special map. The geology of these outer hills was studied practically without a topographic base, a fact in part accounting for the shortcomings of the map.

is located near Goldfield. Wells at Goldfield, in the Siebert lake beds, have struck water, but the main water supply is piped from distant points.

GENERAL GEOLOGY.

In the detailed survey of the special area Mr. Ransome and his assistants recognize the following formations, named from the oldest to the youngest: Cambrian (?) jasperoid; post-Jurassic granite; first andesite; earlier rhyolite; andesite and dacite, the latter formation probably being slightly younger; quartz latite and rhyolite; Siebert lake beds, with associated quartz-basalt and rhyolite flows; later rhyolite; later tuffs, and basalt.

Of these formations the Cambrian jasperoid and the first andesite and rhyolite appear to be confined to the special area and are described in another bulletin of the Survey.[a] The rhyolite flow breccia occurs only in thin flows and, because of the small scale employed, is mapped with the basalt.

SEDIMENTARY ROCKS.

Siebert lake beds.—The Siebert lake beds cover considerable areas in the central and southern portions of the Goldfield hills and form a band between the basalt and the Recent desert gravels. The actual distribution of the formation is greater than that shown on the map, since small areas lie on the eroded surfaces of the andesite and the latite and rhyolite in many places. These lake beds are gray, yellow, or reddish sandstones, shales, grits, and conglomerates, composed predominantly of material derived from the Tertiary lava flows, but subordinately of Paleozoic sedimentary rocks and post-Jurassic granular igneous rocks. The larger bowlders are 2 feet in diameter. Sections on the mesa scarp, in which the base is not exposed, show a vertical exposure of 240 feet. The Siebert lake beds were folded and the edges of the beds truncated by erosion prior to the outflow of the later rhyolite. The formation closely resembles the Siebert lake beds of Tonopah [b] and is presumably of Miocene age.

Later tuffs.—On the mesa scarp to the west and northwest of Goldfield 10 to 20 feet of tuffaceous sandstone lie above the later rhyolite and beneath the basalt. These sediments, probably deposited under water, are characterized by many bowlders of a silvery-gray pumice, presumably derived from the erosion of the surface of the underlying rhyolite. The position of these tuffs beneath the basalt with no marked erosional interval indicates that they are probably to be correlated roughly with the lower portion of the Pliocene-Pleistocene older alluvium occurring elsewhere in the area studied, although they were probably deposited in a lake, the predecessor of the older playas.

[a] Ransome, F. L., Preliminary account of Goldfield, Bullfrog, and other mining districts in southern Nevada, etc.: Bull. U. S. Geol. Survey No 303, 1907, 98 pp., 5 pls.

[b] Spurr, J. E., Prof. Paper U. S. Geol. Survey No. 42, 1905, p. 51.

IGNEOUS ROCKS.

Post-Jurassic granite.—In addition to the granite mapped by Ransome in the special area, pebbles of alaskite and a somewhat less siliceous rock, probably to be correlated with the darker monzonite porphyry of the northeast spur of the Silver Peak Range, occur in the Siebert lake beds.

Andesite and dacite.—Ransome found that andesite flows and dacite intrusions, the latter probably being slightly younger, followed the extrusion of the earlier rhyolite. These formations have been separated in the detailed study of the Goldfield special area, but since the andesite is by far the more abundant rock, the dacite is here grouped with it. It is the most widely distributed formation in the Goldfield hills and appears from a distance to cover a large area in their northeast portion. The later andesite is a rather dark-gray to light-pinkish rock with dense lithoidal groundmass. Phenocrysts, which equal the groundmass in bulk, consist of feldspar and hornblende, augite, and biotite, one of these latter being present in many cases to the almost total exclusion of the others. The dacite, which covers considerable areas within the district shown on the special map, extends eastward beyond its borders, is usually slightly lighter in color than the andesite and contains a few quartz phenocrysts.

The andesite and dacite are younger than the earlier rhyolite and older than the biotite latite and rhyolite. The later andesite is similar in many ways to the later andesite of Tonopah [a] and the two formations may be roughly contemporaneous. The dacite is correlated with that of the Kawich Range. These rocks are probably of early or middle Miocene age.

Quartz latite and rhyolite.—Several outcrops of quartz latite and rhyolite occur in the hills surrounding the Goldfield special area. While some of these may belong to the earlier rhyolite, the majority should certainly be assigned to the quartz latite and rhyolite. Quartz latite outcrops 2 miles south of Preble Mountain. The rock here has a white stony groundmass, slightly exceeded in bulk by the medium-sized phenocrysts of which black mica is most conspicuous, while glassy feldspar striated or unstriated and slightly smoky quartz also occur. Many small ellipsoidal areas of pumiceous latite are present, indicating that the rock was in part a flow breccia. Under the microscope the quartz latite is seen to have a turbid glassy groundmass, with many eddying flow lines. The larger feldspar phenocrysts are andesine and in total bulk the plagioclase phenocrysts slightly exceed those of orthoclase. Orthoclase in particular is zonally built. Quartz has been greatly and biotite slightly corroded by the magma. The magma must have been very viscous prior to the

---

[a] Spurr, J. E., Prof. Paper U. S. Geol. Survey No. 42, 1905, p. 33.

cessation of movement in it, since the phenocrysts are typically broken fragments. Apatite is an accessory mineral.

Rhyolitic glasses that outcrop 3 miles south of east of Diamondfield probably belong to the biotite-latite-rhyolite series. These gray perlitic glasses are poor in phenocrysts, although biotite and less commonly feldspar and quartz are locally present. Flow banding is evidenced by slight alterations in the color and texture of various bands. The perlitic glass grades into spherulitic and dense pinkish-gray glassy rhyolites. The perlitic facies breaks readily into globular fragments along the perlitic parting. The clear glass in some thin sections contains sparse phenocrysts of orthoclase, quartz, and biotite, while in others the phenocrysts are more abundant and oligoclase is also present.

The rhyolite in the southeast corner of the Goldfield hills has highly developed flow banding, and in consequence the rock has a platy structure which wraps knotlike around the phenocrysts. Blackish glasses with numerous phenocrysts also occur. Microscopic examination shows this glass to be in reality a latite with phenocrysts of oligoclase, orthoclase, biotite, quartz, and hornblende.

On the east side of the Goldfield-Lida road, 2½ miles south of Goldfield, is a poorly exposed area of rhyolite. The white lithoidal groundmass, much of it brown or red through iron stains, somewhat exceeds in bulk the abundant phenocrysts. The latter include glassy unstriated feldspar, somewhat less quartz, and a few black mica tablets. Fragments of a pumiceous rhyolite are present. Under the microscope the groundmass appears as a turbid, slightly devitrified glass, showing flow lines and spherulites. A few rather large oligoclase phenocrysts are associated with those of orthoclase, quartz, and biotite. The phenocrysts are all somewhat embayed by magmatic corrosion and were broken into fragments during the flow of the nearly viscous lava. This rhyolite probably belongs to the series under consideration, although it may be one of the later rhyolites. One mile farther south a similar rhyolite outcrops, and this clearly belongs with the quartz latite and rhyolite series, since it was a low boss when the mesa basalt was extruded and that formation flowed in a gentle arch over its surface.

The quartz latite and associated rhyolite are older than the Siebert lake beds and younger than the andesite.

Quartz basalt.—The basalt to the east of the Lida road and 2 miles south of Goldfield is probably a portion of the quartz-basalt flow which Ransome found interbedded with the Siebert lake beds in the special area.

Later rhyolite.—Lying for the most part in a horizontal position upon the eroded Siebert lake beds is a rhyolite flow, which is separated from the overlying basalt by the Pliocene tuffs. This formation is too thin to be shown on the map, but it forms a narrow band from 50 to 100 feet thick beneath the basalt capping of the mesa and occurs also to the south of Goldfield in the area here mapped as Siebert lake beds.

The later rhyolite includes a number of facies of acidic flow rocks. On the Goldfield-Bullfrog road, 3 miles east of south of Goldfield, the formation is 50 feet thick and consists of 4 feet of black glass, above which is 20 feet of faded brick-red dense rhyolite and above this 25 feet of lavender lithoidal rhyolite. The black glass has an excellent flow parting. The middle rhyolite carries rather numerous glassy feldspar phenocrysts, which are, however, subordinate in bulk to the groundmass. It contains many vesicles elongated parallel to the flow, and in these spherulites are developed. The spherulites in some instances form shelf-like partitions in the cavities. Vertical columnar parting is fairly well developed in this portion of the flow. Under the microscope the groundmass, which has many eddying flow lines, shows as a slightly devitrified turbid glass. The phenocrysts are orthoclase, with which are associated a few of plagioclase and bluish-green augite. The presence of considerable quartz in the devitrified glass probably justifies the use of the name rhyolite. The upper member is a flow breccia with lithophysæ here and there throughout its mass. The phenocrysts in the glassy base are seen under the microscope to be orthoclase, bluish-green augite, and quartz. A rhyolite similar to the middle member of this section is exposed 1½ miles southwest of the type locality and north of the Goldfield-Cactus Spring road, 1 mile west of the edge of the alluvial slope. Similar rocks occur beneath the basalt of the mesa west of Goldfield. One facies from this mesa is similar to the upper portion of the flow at the type locality, although phenocrysts, particularly quartz, are more abundant. Apatite is also microscopically visible. Such facies locally include fragments of quartz basalt and the Cambrian jasperoid.

The extrusion of this rhyolite occurred after the folding and erosion of the Siebert lake beds. It is presumably of Pliocene age. Similar rocks occur in the Silver Peak Range, in Slate Ridge, and beneath the basalt of Pahute Mesa.

Basalt.—The youngest formation of the Goldfield hills is the basalt which forms the mesa west of Goldfield, with which the basalt to the northwest of Diamondfield is probably contemporaneous. At one time basalt covered the greater portion of the Goldfield hills.

The basalt is for the most part a dense, compact, dark-gray or black rock, in some instances with phenocrysts equal to the groundmass in bulk and in others without phenocrysts. The phenocrysts, which reach a maximum length of one-fourth inch, include glassy striated laths of feldspar, stocky black columns of augite, and rounded grains of iron-stained glassy olivine. Vesicular facies occur at the top and bottom of the two main flows that compose the 350 feet of basalt. With the introduction of vesicles much of the rock takes on a magenta-red color. The vesicles, which reach a maximum length of 2 inches, are usually elongated parallel to the flow bedding. Some of them are filled by white granular calcite and others by chalcedony. Certain bands contain in a basalt matrix many fragments of basalt, indicating that some portions of the rock solidified before cessation of movement in the magma. Surfaces of flows are often ropy.

Under the microscope the groundmass is usually seen to be holocrystalline and composed of plagioclase laths, augite columns, and magnetite grains. Rarely a little dark glass is also present, and in other thin sections large areas of augite inclose plagioclase laths in the manner of the ophitic texture. A reddish-brown serpentine is developed at the expense of olivine, and a little calcite is formed from plagioclase.

The basalt forms a typical mesa, and buttes isolated in the desert gravels have been carved from it by erosion. This rock is to be correlated with other basic flows widely distributed throughout the area surveyed, which are presumably of late Pliocene or early Pleistocene age.

STRUCTURE.

As Ransome has shown, the Goldfield hills have a domical structure, the oldest formations being confined to the central portion of the hills. He further shows[a] that the older Tertiary rocks have been subjected to considerable folding and some faulting. The Siebert lake beds usually dip away from the central portion of the Goldfield hills. Even the later basalt has been subjected to tilting, in the main away from the center of the dome, and also to some normal faulting. These faults have usually an east-west or north-south strike.

ECONOMIC GEOLOGY.

The ore deposits of Goldfield have been treated at some length in the bulletin already mentioned.[b] Outside of the special area few prospects exist. The andesite and rhyolite 3 miles southeast of Preble Mountain have been altered similarly to those of the productive territory in Goldfield, and the area has been located by prospectors.

ALKALI SPRING FLAT.

The broad valley north of Montezuma Peak is separated from Big Smoky Valley by a gravel-covered divide. The valley has gently sloping sides, except on the west, where it rises rather steeply to the Lone Mountain foothills. The lowest portion, 4,850 feet above sea

---

a Ransome, F. L., Bull. U. S. Geol. Survey No. 303, p. 21.   b Idem, pp. 22-39.

level, is occupied by a playa, situated considerably west of the center of the valley. Inliers of rock do not occur at a distance from the encircling hills, and the recent alluvial deposits are probably deep. The Ramsey and Klondike wells struck water at 240 and 140 feet, respectively. The Ramsey Well is 100 feet higher than the Klondike, and water could probably be obtained near the playa at shallower depths.

SOUTHERN KLONDIKE HILLS.

TOPOGRAPHY AND GEOGRAPHY.

Between Ralston Valley and the flat east of General Thomas Camp are two groups of conical hills and ridges joined by a broad sag formed of Siebert lake beds. These hills are separated from the Goldfield hills by a depression filled with Recent alluvial deposits. The highest peak in the northern part of the hills reaches an altitude of 7,000 feet. The east slope of the group is comparatively gentle, while the west slope is steep. The hills are bare of timber and without water.

GENERAL GEOLOGY.

The formations of the Southern Klondike hills, named in order from the oldest to the youngest, include Cambrian sedimentary rocks, post-Jurassic granite, earlier rhyolite, Siebert lake beds, earlier quartz latite, basalt, later rhyolite, later quartz latite, and earlier alluvium. The mapping to the south of the area of Siebert lake beds is approximately correct, while that to the north is less accurate.

SEDIMENTARY ROCKS.

Cambrian.—Cambrian rocks cover a considerable area in the southern part of these hills. They consist of an interbedded series of limestones, jasperoids, and shales, named in the order of their abundance. Neither the top nor the bottom of the series is exposed, but it is many hundreds of feet thick. The limestone, by far the predominant member of the series, is dark gray or blue black, fine grained, compact, and crystalline. Bedding planes are from 2 to 3 feet apart. Weathered surfaces are blackish gray in color and smooth in contour, although minor irregularities are produced by the varying hardness of the rock. The limestone grades into a silicified facies, best styled a jasperoid. This is ordinarily a dense-banded rock of black and gray color, the laminæ of which are in many cases but one-fourth inch in thickness. It breaks with a conchoidal fracture. The slaty shale is rather fine grained and of dark-gray or greenish-gray color. Even minor lenses and thin parting bands of of fine-grained sandstone are rather unusual. The sediments form hills of moderate slope with numerous exposures. No fossils were found, but the lithology of the series is similar to that of the Lone

Mountain hills to the west, and in consequence the rocks are almost certainly Cambrian.

Siebert lake beds.—The intensely dissected gap between the two groups of hills is composed of white or yellow, well-bedded tufaceous sandstones, conglomerates, and clays, slightly consolidated. About 400 feet of this formation is exposed. Thin sheets of rhyolite appear to be interbedded with the sediments at the south end of the exposures. These soft horizontal beds are eroded into numberless low rounded hills and gullies. Veins of white calcite from one-fourth inch to 1 foot thick cut the sediments and stand in relief on weathering. These beds are similar lithologically to the Siebert lake beds of Tonopah,[a] and like them have thin sheets of the Tonopah rhyolite near their top.

Older alluvium.—North of the large isolated latite cone on the west side of the hills occur small exposures of white sandstones and conglomerates. The pebbles and bowlders consist of latite and biotite-hornblende andesite. These beds are perhaps to be correlated with the older alluvium, probably of late Pliocene or early Pleistocene age.

IGNEOUS ROCKS.

Post-Jurassic granite.—Three small areas of granitoid rocks lie near Southern Klondike. Three-fourths of a mile northwest of the village a granite sheet 250 feet thick, which courses northeast, injects the Cambrian rocks parallel to their bedding. This is a fine-grained rock, although minor portions are of medium to coarse grain. Under the microscope it shows as a rather fresh muscovite granite in which some oligoclase occurs with orthoclase. The quartz shows gentle undulose extinction, but the deformation indicated is not great. A little sericite and less kaolinitic material has been developed in the feldspars. This granite weathers into small rounded bowlders, and exposures are scarce. It is complexly cut by joints 6 inches apart, and in those parallel to the bedding of the surrounding sediments quartz veins occur. A small area of a similar muscovite granite of medium grain is exposed one-fourth mile west of Southern Klondike. An aplitic sheet of the same rock injects the limestone in a mining tunnel at the village. The sedimentary rocks are more or less metamorphosed in the vicinity of the granite, the shale in particular being altered to a knotted silvery schist. Granite cuts the Cambrian rocks and occurs as pebbles in the Siebert lake beds. It is believed to be of post-Jurassic age.

Earlier rhyolite.—Prior to the eruption of the earlier rhyolite which occurs near Southern Klondike the Cambrian rocks and the granite were eroded into a group of hills of slightly greater relief

---

[a] Spurr, J. E., Prof. Paper U. S. Geol. Survey No. 42, 1905, pp. 51–55.

than the present Southern Klondike hills. Since that extrusion erosion has removed considerable masses of the once more extensive rhyolite and partially exhumed the sedimentary rocks. The rhyolite at the northern boundary of the area mapped, on the west side of the hills, is also probably the earlier rhyolite. It occurs in flows and dikes, the flows overlying the Cambrian rocks, the dikes clearly cutting them. The white or pinkish or lilac-white lithoidal groundmass predominates in bulk over the small phenocrysts, among which feldspar is more abundant than quartz. Narrow wavy flow bands wrap around the phenocrysts. One-half mile north of Southern Klondike a resin-colored glassy facies was noted. Rhyolite tuffs and flow breccias are associated with the normal igneous rock near the boundary of the area.

The rhyolite is cut by two sets of joints from 2 to 5 inches apart, and the flow parting constitutes a third plane of weakness. In consequence the residual fragments are rectangular or platy. Weathered fragments are characterized by small pits, the casts of feldspar phenocrysts. The rhyolite forms yellowish or slightly pinkish-white conical hills or low areas. The single slide examined has a brown devitrified groundmass showing flow lines and spherulites. Quartz and orthoclase predominate in the groundmass, while the rather sparse phenocrysts are orthoclase, quartz, and plagioclase. The rock is a rhyolite tending toward latite. The earlier rhyolite is tentatively correlated with the Miocene Tonopah rhyolite-dacite [a] of Tonopah, a correlation already suggested by Spurr [b] for the northern mass.

Earlier quartz latite.—In the northwest portion of the hills, near the areal boundary, is an outcrop of quartz latite of dull-lilac or medium-gray color, which, however, as mapped, may include some basalt. The dense groundmass contains numerous crystals of white striated and unstriated feldspar reaching a maximum length of one-eighth inch, and black mica. Under the microscope the predominant groundmass shows as a devitrified glass containing considerable quartz. Plagioclase and orthoclase are equally abundant phenocrysts, while biotite is in some specimens surrounded by a reaction rim of magnetite. Magnetite, apatite, and zircon are accessories. The latite flow overlies the Siebert lake beds. The rock is rather similar to the later rhyolite of the Goldfield hills, although less siliceous, and the two are perhaps contemporaneous. The Goldfield rhyolite underlies and is older than the basalt.

Basalt.—The large mass of volcanic rock capping the Siebert lake beds northeast of Southern Klondike appears to be an eroded basalt flow, as are the hills between this mass and the large playa crossed by the road from Tonopah to Cactus Spring. It is probable

---

[a] Spurr, J. E., Prof. Paper U. S. Geol. Survey No. 42, 1905, pp. 51–55.
[b] Op. cit., p. 99.

that this basalt is to be correlated with that of Tonopah and the Monitor Hills and that it is of Pliocene age.

Later rhyolite.—The eastern side of the hills on the boundary of the area is formed of rhyolite in flows. The groundmass, much of which is glassy, when fresh is white or light gray, but in places it is stained pink, yellow, or brown by iron oxides. Medium-sized feldspar phenocrysts are more abundant than those of quartz and biotite. This rock is perhaps contemporaneous with the Oddie rhyolite [a] of Tonopah, which is younger than the Siebert lake beds.

Later quartz latite.—The prominent isolated cone between the roads 4 miles west of north of Southern Klondike, several small exposures east of this cone, and a considerable portion of the northern part of the hills east of the Tonopah-Goldfield road are composed of a lilac-gray igneous rock. A black glassy facies occurs on the east side of the cone. As phenocrysts, fresh and altered, feldspars are more abundant than quartz grains, biotite tablets, and hornblende columns. Under the microscope the groundmass appears as a spherulitic glass in which considerable calcite is developed. Orthoclase predominates over plagioclase, which proved in one determination to be an acidic labradorite. Apatite, magnetite, and zircon occur as accessories.

The latite is characterized by vertical flow lines, particularly near the contact with the rhyolite, and the mass is probably a volcanic neck. It is very similar to the Brougher dacite [b] of Tonopah, although somewhat richer in hornblende, and the two are presumably contemporaneous and of Pliocene age.

STRUCTURE.

The Cambrian rocks have been complexly folded, the chief structural feature being an anticline with a northeast-southwest axis. Superimposed upon this anticline are numerous lesser anticlines and synclines, and at many points the beds are intensely crenulated. The rocks are horizontal or dip at varying angles up to 90°, those of 45° predominating. The folding probably immediately preceded and accompanied the granite intrusion. Faults of small displacement occur, and brecciation is common. The sediments are cut in every direction by veins of calcite and less commonly by those of quartz. The veins are not folded, and hence fracturing occurred after the folding of the series and was probably connected with the granitic intrusions. Faulting and some tilting have occurred since the formation of the Tertiary rocks, although it is not known whether the tilting is dependent on faulting or is due to flexing without rupture. On the southern edge of the hills dikes of earlier rhyolite are faulted, but

---

[a] Spurr, J. E., Prof. Paper U. S. Geol. Survey No. 42, 1905, pp. 49–50.
[b] Ibid., pp. 44–48.

lack of time did not permit their being mapped. Some faulting also occurred after the formation of the ore deposits in the northern part of the hills. The basalt northeast of Southern Klondike dips to the south.

ECONOMIC GEOLOGY.

GOLD AND SILVER.

Gold and silver ores were discovered in the Cambrian rocks at Southern Klondike in March, 1899, by J. G. Court and T. J. Bell. Since that time the prospects have been worked more or less continuously. It was while on a trip to these prospects in 1900 that J. L. Butler discovered the veins at Tonopah. From one group of claims shipments of ore ranging in value from $200 to $284 per ton and aggregating $50,000 are reported. Several thousand feet of tunnels, shafts, and inclines have been driven. In June, 1905, eight men were at work in the camp.

The ore deposits are of three kinds—first, quartz veins [a] which are parallel to the bedding of the Cambrian rocks, and which carry predominant silver values; second, veins along the contact of the sedimentary rocks and rhyolite dikes, the values, about $45 per ton, being predominantly gold, and third, thin veins of quartz carrying silver-bearing galena and cerussite in granite along joint fractures parallel to the bedding of the surrounding Cambrian rocks. The veins along contacts apparently show post-mineral faulting.

The quartz veins parallel to the bedding planes, which occur at and 1 mile east of the village of Southern Klondike, are of greatest value and interest. These are tabular lenses of quartz, from a few inches to a foot or more in thickness. Horses of limestone are included. Adjacent veins connected by cross veins of quartz or completely separated by thin bands of limestone form in some instances mineralized zones 14 feet thick. Quartz appears to have filled a fissure in the limestone, brecciation of the limestone having accompanied the fissuring. The contact between the limestone and quartz is sharp, and important silicification of the limestone has not occurred. Vugs lined with acicular quartz crystals are rather characteristic.

The original sulphides deposited simultaneously with the quartz and disseminated in small masses in it are, in the order of their abundance, galena, copper sulphide,[b] and iron pyrites. The secondary ores include cerargyrite, chrysocolla with less malachite and azurite, specular hematite, and cerussite in brownish granular masses and to a less extent in crystals. These secondary minerals surround the sulphides

---

[a] For a detailed description of this type see Spurr, J. E., Economic Geology, vol. 1, 1906, pp. 360–382.

[b] Stetefeldtite, according to J. E. Spurr, Trans. Am. Inst. Min. Eng., 1905, p. 961.

and fill cavities and cracks in the quartz. Some calcite and gypsum, accompanied by sulphur, are secondary gangue minerals. Sulphur in places forms masses 3 inches in diameter. Assays did not detect the presence of gold at the surface, although gold values are encountered 50 feet below the surface in one tunnel. One small aplite dike was noted near a quartz vein, and masses of quartz rhyolite are not far distant, although their proximity has not apparently determined an unusual abundance of quartz veins. The prospects are yet well above water level.

In Southern Klondike fissures and brecciated zones were opened in the Cambrian sediments along bedding planes. Water filled these cavities with silica and lead, copper, and iron sulphides carrying silver and less gold. The quartz veins have since been fractured and faulted, and surface waters have developed a number of secondary minerals. In this district the presence of quartz veins at the contact of rhyolite and limestone shows the gold veins to be of late Tertiary age, but the more important silver veins are presumably of post-Jurassic and pre-Tertiary age.

Water is obtained at the Klondike Well, 4 miles distant. The Southern Klondike hills are bare of timber. When the district was visited Tonopah was the supply and shipping point.

The earlier rhyolite northwest of the Tonopah-Goldfield road near the boundary of the area mapped has locally been silicified and kaolinized by mineralizing waters, and in it some quartz veins occur. The reported values are in gold with less silver. The Kankakee Mining Company has a 100-foot shaft, which passes through rhyolite, tuffs, and breccia. Dense quartz fills the interstices of breccia and also occurs in veins. The ore is free milling and is said to run about $30 in gold and silver per ton, although assay returns have reached $317. Spurr[a] says, concerning these veins: "Through these rhyolites run strong and persistent veins of quartz and delicately colored chalcedony veins, sometimes containing pyrite. In some parts of some of these veins, especially in the oxidized portions, rich assays have been obtained." In his report on Tonopah Spurr correlates these veins with those of the Tonopah rhyolite-dacite.[b]

IRON ORES.

A several places northwest of Southern Klondike lenses of porous hematite occur in limestone. These masses occupy joints in the limestone and are from 5 to 10 feet wide and from 50 to 100 feet long. The hematite appears to be partially a replacement of limestone and partially a cavity filling. Cubes of pyrite are present in the neighboring limestone. These masses of iron ore are probably gossan and with depth will pass into pyrite veins.

[a] Spurr, J. E., Bull. U. S. Geol. Survey No. 213, 1903, p. 87.
[b] Spurr, J. E., Prof. Paper U. S. Geol. Survey No. 42, 1905, p. 99,

RALSTON VALLEY.

The south end of Ralston Valley is within the area surveyed and lies between the Monitor, Goldfield, and Southern Klondike hills and the Cactus Range. Hills surrounded by gravels extend from the Cactus Range northeastward toward the Kawich Range and separate Ralston Valley from Cactus Flat, and on the divide the Recent gravel deposits are probably but a few hundred feet thick. Ralston Valley is very shallow, the rim of hills being in many places but 200 feet above the playa (elevation, 5,270 feet). Low sand dunes lie on the northeast of the playa. A well sunk in the playa struck water at a depth of 240 feet.

STONEWALL FLAT.

Stonewall Flat, of northeast-southwest trend, is inclosed by the Cactus Range, Stonewall Mountain, the Goldfield hills, and the Mount Jackson hills. A gently sloping detrital divide with rock inliers separates it from Ralston Valley to the north, while the detrital barrier between it and the valley north of Slate Ridge is less than 100 feet high. Several small playas, between 4,700 and 4,800 feet above sea level, occupy minor depressions in the valley. The playas nearest the Cactus Range are encircled by low sand dunes. A well on the Goldfield-Gold Crater road, 2 miles east of the largest playa, struck water at a depth of about 110 feet. The water contains small amounts of salt and probably sodium carbonate. On the west side of the largest playa are low dissected bosses of dazzling white older alluvium, the highest of which are 25 feet above the present playa. Massive beds of white or cream-colored clay of fine texture, with thin interbedded layers of white limestone, are exposed. A number of compact lenticular lime-carbonate concretions, with hackly surface, which reach a maximum length of 6 inches, are embedded in the clay. These beds resemble those of Gold Flat and the valley north of Slate Ridge and have suffered about the same amount of erosion. They were evidently deposited in an ancient playa, possibly of late Pliocene, but probably of early Pleistocene, age.

STONEWALL MOUNTAIN.

TOPOGRAPHY AND GEOGRAPHY.

Stonewall Mountain rises from the northwest border of Pahute Mesa. It was named, according to some, from Gen. Stonewall Jackson, but by others the name is said to have been derived from the precipitous, wall-like northern face of the mountain. The rugged mountain group, 9 miles in diameter, is formed of sharp peaks and steep ridges culminating in a summit 8,390 feet high. Stonewall Mountain is an excellent example of the symmetrical erosion of a homogeneous mass of an approximately circular horizontal plan. Deep canyons radiate from the central peak, ending at the alluvial slopes 2,500 to 3,000 feet below. The most striking feature of the mountain is the fault scarp at Stonewall Spring, a sheer wall 800 to 1,000 feet high. Erosion has cut partially through this scarp, forming a narrow V-shaped gap, the lowest point of which is at least 100 feet above the stream bed to the south. Behind this wall the valley is broad and U-shaped, regaining its V shape some distance back in the mountains. Other streams near by are characterized by narrow canyons, which open upstream into more mature valleys.

The mountains above an elevation of 6,100 feet are clothed with a sparse although locally heavy growth of juniper, piñon, and mountain mahogany. Excellent pasturage covers the higher peaks and valleys. Groves of the tree yucca occur on alluvial slopes to the west and on Pahute Mesa to the southeast. Stonewall Spring yields daily about 3,000 gallons of pure cold water. A stream 100 yards long flows from a spring in the gulch one-half mile east of Stonewall Spring. Another spring is located 1 mile west of the culminating peak of the range.

GENERAL GEOLOGY.

The formations of Stonewall Mountain are, in ascending order, Cambrian limestone, post-Jurassic granitoid igneous rocks, earlier rhyolite, quartz syenite and quartz-monzonite porphyry, Siebert lake beds, later rhyolite, and basalt.

SEDIMENTARY ROCKS.

Cambrian.—Three small masses of dark-gray crystalline limestone, cut by many white calcite veinlets, protrude from the igneous rocks near the north edge of the mountain. Angular inclusions of limestone and jaspilite are embedded in the older rhyolite, the monzonite porphyry, and the basalt. The limestone is similar lithologically to that of the Cuprite mining district, and, like it, is probably of Cambrian age. In contact with quartz-monzonite porphyry, near the northwest boundary of the mountain mass it has been metamorphosed to a white marble containing epidote.

Siebert lake beds.—Incoherent, well-bedded tuffaceous sandstones and conglomerates of red, yellow, white, and greenish-white color cover a small area east of Stonewall Spring. The bowlders, which are well rounded to semiangular, are largely of the earlier rhyolite. A thickness of about 500 feet is exposed. These beds are without much doubt the equivalent of the Siebert lake beds.

IGNEOUS ROCKS.

Post-Jurassic granite and granite porphyry.—A few inclusions of granite occur in the earlier rhyolite and prior to the extrusion of that rock granite may have been exposed in these mountains. A narrow dike of granite porphyry cuts the easternmost limestone. This rock has a finely granular yellowish groundmass in which are inclosed medium-sized quartz and feldspar phenocrysts. In the field this resembles the granite porphyry near Alkali Spring in the Silver Peak Range and it may be genetically related to the post-Jurassic granites. Angular fragments of a gray monzonite porphyry occur in the earlier rhyolite near the summit of the group and these may be correlated with similar rocks of the Silver Peak Range that are also related in origin to the post-Jurassic granites.

Earlier rhyolite.—The most widely distributed formation of Stonewall Mountain is a rhyolite characterized by many small colorless and glassy or white and opaque unstriated feldspar phenocrysts. Near Stonewall Spring the rhyolite is medium gray and compact; that of the highest peak is also compact and of a faded red color; the rhyolite of the small hill in the detrital apron on the southwest side of the mountain is a gray lithoidal rock. The rhyolite from the last two localities contains biotite phenocrysts.

An indistinct original banding is locally present and some facies are a flow breccia. The rhyolite is probably a flow, although the vertical position of the banding in places and the texture of some facies indicate that the rock may possibly be an intrusive mass. The rhyolite as exposed at present is several thousand feet thick, and the granitoid habit of the quartz syenite which injects it suggests that it was probably once much thicker. The rhyolite has well-developed joints over large areas which express themselves in the topography as straight elements. Where the joints are less closely spaced the rhyolite weathers in rounded forms like a granite.

Microscopic examination shows the rhyolite to have either a devitrified glassy or a microgranitic groundmass composed of predominant orthoclase with fewer quartz grains and many small magnetite cubes. The groundmass varies widely in grain from one part of the thin section to another, but is everywhere rather fine grained. Orthoclase phenocrysts, many of them microperthitic, with ragged borders which inclose some of the groundmass anhedra, are abundant. Rarely plagioclase laths, well-formed biotites, very ragged hornblende, and in some thin sections one or more quartz phenocrysts are associated. Zircon and apatite are accessories.

The mature topography of Stonewall Mountain shows that the rhyolite is old, and pebbles of it occur in the Siebert lake beds. It is believed to be older than the rhyolite of the Kawich Range, with which it has no lithologic affinities, and is probably of early Eocene or even late Cretaceous age.

Quartz syenite and quartz-monzonite porphyry.—Dikes and elongated masses of quartz syenite and quartz-monzonite porphyry, which grade into one another and are hence contemporaneous, intrude the earlier rhyolite at widely separated localities in the mountain. Dikes of quartz syenite less than 200 feet wide intrude the earlier rhyolite on the fault scarp just east of Stonewall Spring. The quartz syenite is a grayish-white, medium-grained granitoid rock composed of unstriated feldspar with a little biotite. Microscopic examination shows the presence of a few small wedges of quartz between the feldspar grains and the biotite tablets. Ilmenite, apatite, and zircon are accessories. The granitoid habit of these small masses indicates that its magma solidified beneath a considerable depth of rhyolite now removed by erosion.

The quartz-monzonite porphyry is typically porphyritic in habit, with a gray and black speckled holocrystalline groundmass. The phenocrysts, which predominate over the groundmass, include white or gray glassy feldspar, striated or unstriated, and black or bronze-brown biotite. The largest phenocrysts are one-fourth inch in length. The contact facies of the dike three-fourths of a mile southeast of the summit of Stonewall Mountain is of slightly finer grain and includes many fragments of the earlier rhyolite. The rock weathers into rounded bosses and disintegrates into spheroidal bowlders stained brown or reddish by iron oxides. The microscope shows that the microgranitic groundmass is composed of orthoclase grains, many of them twinned according to the Carlsbad law, with a few associated quartz wedges and biotite shreds. The phenocrysts include both plagioclase, in large complex individuals, and orthoclase. Phenocrysts of biotite and smaller ones of light-green augite are also present. Magnetite, apatite, and zircon are the accessory minerals. Calcite, sericite, and kaolin are present as alteration products of the feldspars.

One and three-fourths miles upstream from the easternmost area of Cambrian limestone a poorly exposed black glass with rather numerous feldspar phenocrysts apparently cuts the earlier rhyolite. The exposure is too small to map. Under the microscope this proves to be a glass with orthoclase and a few bluish-green augite phenocrysts. It may well be a glassy form of the quartz syenite. A small fragment of hornblende andesite, unlike any rock exposed in the mountain, is included in the thin section.

The quartz syenite and quartz-monzonite porphyry of Stonewall Mountain intrude the earlier rhyolite, and pebbles are embedded in the Siebert lake beds. They most closely resemble the post-Jurassic monzonite porphyry of the Silver Peak Range (see p. 59), although they are not very different from the monzonite porphyry of the Kawich Range and their correlation with the latter is more nearly correct. The rocks are believed to be of late Eocene age.

Later rhyolite.—The later rhyolite forms the low, massive brown and gray domes of the eastern part of Stonewall Mountain. Two outlying hills on Pahute Mesa to the south are of the same formation. The later rhyolite has a dense gray stony to glassy groundmass. In it are numerous medium-sized glassy unstriated feldspar and fewer rounded quartz phenocrysts. The feldspar crystals often exhibit beautiful plays of color, purples and blues being refracted from numerous cracks. The glassy facies locally show well-developed flow lines and spherulites, while slightly vesicular facies also occur. The later rhyolite breaks into platy fragments and in places a granite-like weathering is developed. The rhyolite flow, as now exposed, 1,200 feet thick, is in broad, gentle flexures which may be due in part to the slightly uneven surface upon which it flowed. Microscopic examination shows a turbid glass in which orthoclase and quartz and fewer acidic plagioclase phenocrysts are associated. A little magnetite is also present.

For a distance of 2 miles west of the rhyolite, pebbles of black glassy obsidian are common upon the surface of Stonewall Mountain. These pebbles may indicate that the rhyolite once extended some distance farther west, although similar rocks were not noted in the later rhyolite section.

The later rhyolite overlies and is in consequence younger than the Siebert lake beds. The contact between the rhyolite and the basalt which forms the top of Pahute Mesa at the northeast corner of Stonewall Mountain was examined at one point. The contact is slightly undulating, and the base of the overlying rhyolite for 5 to 6 feet is a flow breccia of massive blocks of vesicular rhyolite. This apparently indicates that the rhyolite is younger than the basalt beneath it. The topography of the rhyolite, however, is so much more mature than that of the Pahute Mesa basalt that it is believed that the rock here underlying the rhyolite is a portion of an older basalt which extended from Stonewall Mountain as a shelf and which happened to be on a level with Pahute Mesa. The later rhyolite is probably to be approximately correlated with the later rhyolite of the Southern Klondike hills.

Basalt.—A small mass of a grayish-black basalt outcrops 1 mile east of south of Stonewall Spring. Basalt breccias indicate that it is a flow. The age of this rock is unknown, but from the apparent absence of its bowlders in the Siebert lake beds it is probably to be correlated with the later basalt of the Goldfield hills.

STRUCTURE.

The Cambrian limestone is practically horizontal, the folding being broad and open like that of the hills of the Cuprite mining district. In restricted areas, however, the rock is closely folded. Prior to the extrusion of the earlier rhyolite the Paleozoic rocks appear to have had a gently accentuated topography. The most striking structural and topographic feature of Stonewall Mountain is the fault on the front of the mountain group, near Stonewall Spring. This fault strikes N. 65° E. and dips 70° N. Minor faults and sheeting parallel to the main fault occur for a distance of one-half mile south of the mountain front. The Siebert lake beds strike east and west and dip 5° to 50° N. This fact probably indicates that Stonewall Mountain is on the upthrown side of the fault. The fault fissure was healed by a quartz vein and has been reopened several times and again healed. (See p. 88.) Since the fissure filling erosion has uncovered the fault and a secondary fault scarp, whose position is determined by the resistant quartz vein, has been formed. The topographic form of the valley of the Stonewall Spring drainage line (see p. 84) indicates that in comparatively recent time the mountains behind the fault line have been uplifted. Fig. 7 is a section through the north side of Stonewall Mountain.

ECONOMIC GEOLOGY.

Quartz veins and stringers filling faults, joints, and the cavities of brecciation are very abundant in the earlier rhyolite, quartz-monzonite porphyry, and quartz syenite near Stonewall Spring. The most prominent vein follows the fault scarp immediately south of the spring. The quartz vein is in some places simple and 40 feet wide; in others it is complex and composed of many parallel veins. This vein or other veins of approximately similar strike extend 2 miles west and 1 mile east of the spring. The quartz is white or colorless, rarely greenish yellow, and is beautifully crustified. Vugs with quartz crystals or mammillary quartz are common. Movement reoccurred along this fault and the quartz has locally been fractured and slightly displaced, later quartz filling the cavities. Heavy stains of limonite and slight stains of azurite were noted in the quartz, and pyrite is locally developed. Prospectors report from a trace to $6 in gold per ton at a number of places on this vein. Similar but smaller quartz veins occur throughout the areas of earlier rhyolite and quartz-monzonite porphyry.

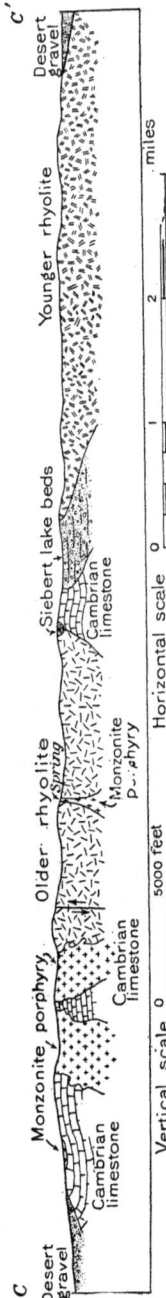

Fig. 7.—Section across north face of Stonewall Mountain, through Stonewall Spring.

A prospect is located in the Cambrian limestone area 2 miles east of Stonewall Spring. A thin vein of quartz cuts a granite-porphyry dike, which here intrudes limestone. The quartz is slightly limonite stained and said to carry from $2 to $56 in gold per ton.

RANGES NORTH OF PAHUTE MESA.

To the north of Pahute Mesa, which occupies the central portion of the area surveyed, lie the Cactus, Kawich, Reveille, and Belted ranges. These, with the exception of the Cactus Range, trend north and south. They have as their predominant formation a rhyolite of early Miocene age.

CACTUS RANGE.

TOPOGRAPHY AND GEOGRAPHY.

The Cactus Range lies between Stonewall and Cactus flats. To the north its outlying hills are separated from the Monitor Hills by Ralston Valley; its low southern portion is buried beneath lava flows of Pahute Mesa. The range as defined is 22 miles long and has a crest line coursing northwest. The low central part is almost cut in two by gently sloping valleys filled with alluvial material. To the north of this median line the range culminates in Cactus Peak, a symmetrical cone 7,550 feet high and a landmark visible for many miles. The highest peak of the range south of the center line is a black massive mountain 7,600 feet high.

Piñon and juniper grow sparsely on the higher parts of the range south of the median line, and the tree yucca abounds on the lower hills and upper alluvial slopes. Good grazing is found on the alluvial slopes next to the mountains and in some of the valleys. Cactus Spring has a daily flow of about 500 gallons of clear cold water. The water of Alkali Spring is cool, but slightly saline. Several other small water holes are reported to the west and north of Alkali Spring. Antelope Spring flows from 300 to 400 gallons of cool palatable water daily, and two smaller evanescent springs are situated in gulches to the south within one-half mile of Antelope Spring.

GENERAL GEOLOGY.

The succession of formations exposed in this range, from the base up, is as follows: Pogonip limestone, Eureka quartzite, Weber conglomerate, granite, diorite porphyry, hornblende-biotite latite, earlier rhyolite, biotite andesite, augite andesite, later tuffs (?), later rhyolite (?), and basalt.

SEDIMENTARY ROCKS.

Pogonip limestone.—A small exposure of dark-gray fine-grained limestone, surrounded by alluvial deposits, occurs in the broad valley 3 miles south of the Goldfield-Cactus Spring road. This limestone is cut by numerous small white calcite veins. On lithologic grounds it is probably the Pogonip limestone of Ordovician age. Bowlders of similar limestone are embedded in rhyolitic tuffs near by and fragments of limestone and jasperoid are inclosed in the rhyolite at several places and in a granite porphyry in the northern part of the range. These fragments were probably derived from Ordovician or Cambrian limestones.

Eureka quartzite.—Two miles southwest of Antelope Spring an area of quartzite lies unconformably below the surrounding rhyolite. This is a fine- to medium-grained quartzite of white, yellow, or red color and is cut by small stringers of white quartz. From its lithologic character it is considered to be the Eureka quartzite (Ordovician).

Weber conglomerate.—Three miles west of south of Cactus Spring a mass of rusty-looking sedimentary rocks, one-half mile in diameter, protrudes through rhyolite. About 150 feet of conglomeratic beds, with the matrix of sand and small pebbles thoroughly cemented, are exposed. Embedded in this matrix are many beautifully rounded pebbles of green and black flint and jasperoid, white quartzite, and black limestone. The roundness of the pebbles, the largest of which are 4 inches in diameter, indicates a long period of attrition and the conglomerate is probably of marine origin. Prior to their inclusion in the conglomerates the pebbles, probably derived from Cambrian, Ordovician, and Silurian rocks, were cut by quartz and calcite veinlets and the limestone silicified to jasperoid. The conglomerate is thus much younger than the Pogonip limestone, but since it contains no granite or diorite pebbles it is believed that it is of Carboniferous age and that it is to be correlated with the Weber conglomerate of the Belted Range.

IGNEOUS ROCKS.

Post-Jurassic granite porphyry and granite.—Midway between Cactus Peak and Cactus Spring is a small area of granite porphyry that is poorly exposed, but lithologically distinct from the surrounding rhyolite. The pinkish-gray rock is of well-developed porphyritic habit, with a finely crystalline groundmass. The phenocrysts, which exceed the groundmass in bulk, are pink glassy feldspar tablets up to three-fourths of an inch in length, a few smaller quartz grains, and fairly abundant biotite flakes. In it are fragments of Paleozoic sedimentary rocks. Under the microscope the groundmass is seen to be a microgranitic mosaic of orthoclase and quartz with here and there a little plagioclase and biotite. The orthoclase phenocrysts, many of which have zonal structure, are in places twinned according to the Carlsbad law. Quartz, biotite, and a few plagioclase phenocrysts are associated. Magnetite and zircon occur as accessory minerals. Both quartz and orthoclase phenocrysts show undulose extinction, which in certain instances is rather strongly developed. The plagioclase phenocrysts and the feldspars of the groundmass are turbid through kaolinization. This granite porphyry closely resembles some of those of post-Jurassic age. Inclusions of a siliceous granite occur in rhyolite at a number of places.

Pre-Tertiary diorite porphyry.—Two miles north of Antelope Spring is a small area of greenish-gray diorite porphyry. The rock has a well-developed porphyritic texture; small gray striated feldspars, grayish-green altered hornblendes, and much smaller black micas lie in a fine-grained gray groundmass. Under the microscope the groundmass appears as a fine microgranitic mosaic of plagioclase and some orthoclase. Of the phenocrysts already mentioned both plagioclase and hornblende are much altered, the plagioclase being sprinkled with epidote, calcite, and zoisite, and the hornblende being more or less completely replaced by epidote, chlorite (ripidolite, in part), and calcite. Ilmenite and apatite are accessory minerals. This rock is practically identical with the older pre-Tertiary diorite porphyry of the Lone Mountain foothills.

Hornblende-biotite latite.—On the west side of the Cactus Range, north of Wellington and south of the median line of the range, low rounded bosses of a much altered greenish-gray rock protrude from beneath the younger rhyolite. The dull groundmass contains biotite plates, white or pale-green areas, apparently altered feldspar, and dark-green areas, probably altered hornblende, while the weathered surfaces show numerous casts of these phenocrysts. The rock contains many well-rounded pebbles, some of grayish-white quartzite (Eureka?), others of the diorite porphyry last described. The largest pebbles are 3 inches in diameter. The latite appears to have flowed out upon an old erosional surface covered by, well-rounded pebbles. Under the microscope the groundmass shows as a glass, now, however, much altered and composed of epidote, calcite, quartz, and orthoclase. The plagioclase phenocrysts are almost completely altered to epidote and calcite, with less chlorite, quartz, and zoisite. The hornblende phenocrysts are altered to the same minerals, although zoisite is as a rule absent. Biotite is replaced by chlorite in association with sagenitic webs of rutile. The phenocrysts and groundmass have thus been altered similarly, although the secondary minerals of the phenocrysts are coarser in grain than those of the groundmass.

Wherever the contact with the rhyolite was seen the latite appears to be the older, a view supported by the absence of rhyolite pebbles in it and by the intense alteration and deformation which it has suffered. Petrographically the latite is rather similar to the later andesite of Tonopah,[a] but mineralogically it is more closely allied to the

---

[a] Spurr, J. E., Prof. Paper U. S. Geol. Survey No. 42, 1905, p. 33.

monzonite porphyry of the Kawich Range; it is therefore tentatively considered the effusive equivalent of that formation and is thus probably of Eocene age.

Earlier rhyolite.—The most widespread formation of the Cactus Range is a rhyolite which occurs in flows. Throughout the range it appears to bear similar relations to the other Tertiary rocks and in the main probably represents a single period of rhyolitic volcanism. More detailed work may, however, prove that some of the rhyolites near Cactus Peak are younger than those of the central and southern portions of the range.

The rhyolites vary among themselves in color, in character of groundmass, and in the relative abundance of the various phenocrysts. The predominant type is a white or gray rock of lithoidal or glassy groundmass, in which are embedded abundant medium-sized, slightly smoky quartz and glassy orthoclase phenocrysts; biotite phenocrysts are small and inconspicuous or absent. Other phases of the rhyolite are black, purple, or red in color. The phenocrysts of quartz are typically corroded grains, although some exhibit the dihexagonal pyramid and prism. Wavy flow bands of slightly different color traverse the groundmass. In many beds irregular fragments of rhyolite are inclosed in a matrix of similar rhyolite, showing that portions of the magma were solidified prior to cessation of movement in the flow. The presence of well-rounded pebbles of Paleozoic rocks in the basal portions of the rhyolite indicates that the surface upon which the lava flowed was covered by such pebbles.

Microscopic examination of several thin sections shows these rocks to be normal rhyolites. The groundmass is a brown glass, and many of the phenocrysts are fractured by flow. One or two acidic plagioclase phenocrysts are present in some sections. Biotite in some instances is altered to chlorite and epidote or muscovite. Accessories are rare, although apatite and ilmenite occur.

On weathering the feldspar phenocrysts are removed and the quartz protrudes slightly. In some portions of the range the rock has the smooth contours of weathered granite, but in the vicinity of Cactus Peak the greater resistance to erosion of certain bands gives a bedded aspect to the series.

Vertical columnar parting is well developed through cooling on Cactus Peak, while on the Goldfield-Cactus Spring road horizontal hexagonal joints occur. The rhyolite is so similar to that of the Kawich, Reveille, and Belted ranges that it is considered to be also of early Miocene age.

Four areas of slightly consolidated rhyolitic sandstones and conglomerates occur in the Cactus Range. Two are on the northwest border of the range near the Cactus Spring-Tonopah road, a third is situated 2 miles southwest of Cactus Spring, and in a fourth area these rocks underlie the augite andesite on the east side of the range. The tuffaceous sandstone, which is well bedded and white or greenish in color, is but slightly consolidated and breaks down readily into a deep sandy soil. Interbedded with the sandstones are conglomerates which, in the area southwest of Cactus Spring, contain bowlders of Paleozoic limestone 3 or 4 feet in diameter. These beds are tentatively considered rhyolitic tuffs deposited in local basins during the rhyolitic extrusion. It is recognized, however, that these sediments may in reality be the Siebert lake beds, in which case the earlier rhyolite of Cactus Range is of late Miocene age and is to be correlated with the later rhyolite of the Belted Range.

Biotite andesite.—Dikes and flows of biotite andesite are widely distributed in the southern part of the Cactus Range, but do not appear to extend far north of the Goldfield-Cactus Spring road. In dikes and irregular intrusive masses and possibly in flows this rock covers considerable areas southeast of Cactus Spring, apparent dikes cut the rhyolite at Wellington, and an andesite flow caps the highest mountain in the southern half of the range. Other areas of biotite andesite occur northeast of Antelope Springs, and a traverse from Wellington to Antelope Springs crossed numerous areas of this rock too small to show on the present map. Dikes of biotite andesite occur on both sides of the Goldfield-Cactus Spring road, but these also are too small to indicate on the map.

The biotite andesite is everywhere more or less altered. The freshest rocks have a dark-gray groundmass in which are embedded abundant medium-sized phenocrysts. Striated feldspars of white color and in many cases of zonal growth are more conspicuous than the altered grayish-green biotite and hornblendelike mineral. More altered facies are greenish gray or purplish red in color. The andesite breaks into sharp joint blocks which on further alteration develop spheroidal weathering. Under the microscope this rock appears to have had originally a pilotaxitic groundmass. Plagioclase phenocrysts are common and grade in size into the laths of the groundmass. One or more orthoclase phenocrysts are present in the majority of slides. Biotite is now a pseudomorph of chlorite, calcite, epidote, quartz, and sagenitic rutile. Aggregates of chlorite, epidote, calcite, and quartz surrounded by reaction rims of magnetite appear in some cases to have the form of a pyroxene and in others that of an amphibole. Probably both minerals were originally present. These pseudomorphs, like the plagioclase phenocrysts, grade into groundmass microlites of similar form. Magnetite is a common accessory.

The biotite andesite cuts the rhyolite in dikes and caps it in flows; inclusions of rhyolite occur in the andesite, and near some of the rhyolite masses the phenocrysts of the andesite become smaller. The andesite, then, is younger than the rhyolite, and from its altered condition it is believed to be older than the augite andesite and basalt. It closely resembles the andesite of the Goldfield hills and may be contemporaneous with it. If so, it is probably of early or middle Miocene age.

Augite andesite.—A flow of dark-gray andesitic rock overlies the tuffaceous facies of the earlier rhyolite 2 miles north of Antelope Springs. In the dense groundmass are blackish-green pyroxene and amphibole columns which reach a maximum length of one-eighth inch. The rock breaks into sharply jointed blocks, in the interstices of which some epidote has developed. Microscopic examination shows that this is an augite andesite with glassy groundmass in which are numerous plagioclase laths, pyroxene crystals, and magnetite grains. Augite phenocrysts with slight zonal structure and twins parallel to the orthopinacoid are abundant. The augite is remarkably fresh, although a little secondary epidote and chlorite is locally present. A few brown hornblende phenocrysts, some of them outlined by a reaction rim of magnetite, also occur. Apatite and magnetite are present as accessory minerals.

The augite andesite forms a flow which is apparently younger than the rhyolite and, to judge from its fresh condition, is probably also younger than the biotite andesite. In the Great Basin the pyroxene andesites are of late Pliocene-Pleistocene age,[a] and their formation usually immediately preceded that of the later basalts.

Later rhyolite (?).—One mile northeast of Cactus Peak are some low hills of purplish-gray rock with rather large feldspar crystals. Tuffaceous beds underlie the igneous rock, and the two rocks are probably to be correlated with the younger tuff and the youngest rhyolite of the Goldfield hills. It is by no means impossible that the same series underlies the basalt in the west slope of the range.

Basalt.—The hill 3 miles southeast of Antelope Springs is composed of rhyolite apparently overlain by basalt, and several hills along the edge of the range on the Cactus Spring-Silverbow road are composed of a similar rock. The dissected mesa slopes on the west side of the range north of Wellington appear from a distance to be a northward extension of the basaltic rocks of Pahute Mesa. Probably to be correlated with the basalt is a reddish-brown vesicular rock which caps a low dome 1½ miles north of Cactus Peak. These basalts appear to overlie the rhyolite and they are probably of late Pliocene or early Pleistocene age.

STRUCTURE.

The Cactus Range is predominantly formed of Tertiary rocks unconformably overlying Paleozoic sedimentary rocks and granites and diorite porphyries probably of post-Jurassic age. The small areas of

---

[a] Spurr, J. E., Trans. Am. Inst. Min. Eng., vol. 33, 1903, p. 333.

Paleozoic sedimentary rocks are gently folded in a manner comparable probably with the folding of the Stonewall Mountain Paleozoic rocks. Minor normal faults were observed in all the Tertiary lavas, although they are much more abundant in the rhyolite and other older formations than in the younger. The rhyolite in places is tilted at an angle of 30°, but it was not determined whether the tilting is due to faulting or to actual folding.

ECONOMIC GEOLOGY.

Wellington, formerly called O'Briens Camp, is situated on low rounded hills in the southwestern portion of the range, 11 miles south of Cactus Spring. Claims were first located in August, 1904, and when visited (July, 1905) several men were doing development work.

The country rock, the earlier rhyolite, is considerably kaolinized and silicified in the vicinity of the veins and is heavily stained by limonite. The rhyolite is apparently cut by dikes of altered biotite andesite of purple color. Both rocks along a zone striking N. 70° E. are cut by quartz veins, many of which strike parallel to the extension of the zone and dip northward. The larger veins are from 2 to 4 feet in width. Connecting these are numerous quartz stringers, which course in all directions, in many places cementing crushed portions of the rock. The quartz is semitransparent, crystalline, and for the most part white, although locally intensely stained by limonite and manganese dioxide. Vugs with small quartz crystals are very common, as is also crustification. Minor veins of calcite were observed. Differential movement has occurred parallel to some of the veins, and much of the quartz is intensely brecciated, while minor faulting across the strike was noted in several places. Microscopic examination shows that the brecciated fragments were first rimmed by fringes of quartz, the interstices being later filled by calcite and limonite. The values reported are largely gold, silver constituting but one-twentieth of the assay value. The ore is free-milling and the gold is in close association with limonite. The quartz and the contained ores were deposited in joints in the interstices of breccia and along small and possibly large fault fissures. The veins have been faulted and the quartz crushed. Only ores oxidized by surface waters have as yet been encountered.

The Cactus Mining Company has a shaft in silicified rhyolite three-fourths of a mile south of Cactus Spring. Coarsely crystalline white quartz veins, with many vugs, cut the rhyolite, and these on surface outcrops are heavily stained by iron compounds. Pyrite and chalcopyrite are sparingly present in the quartz and to a less degree impregnate the surrounding rhyolite. Free gold is reported. Both fissure filling and replacement of the country rock have occurred. To the north of Alkali Spring some quartz veins cut iron-stained kaolinized and silicified rhyolite. Prospectors report that the quartz carries low-grade values.

Over a considerable area south of the Goldfield-Cactus Spring road, beneath the north end of the andesite flow, which caps the high mountain south of the median line, and in several smaller areas the rhyolite has been silicified and kaolinized. (See fig. 4, p. 43.) Quartz veinlets occur in some of these areas. Masses of hematite and limonite outcropping 2 miles S. 60° E. of Cactus Spring may be the gossan of a pyrite vein. Malachite stains occur on joint surfaces in the biotite andesite one-half mile southeast of Cactus Spring.

HILLS BETWEEN THE CACTUS AND KAWICH RANGES.

TOPOGRAPHY AND GEOGRAPHY.

Lying between the Cactus and Kawich ranges is a rather broad ridge of north-south trend, from which low domes arise. The southern part of the ridge is a mesa sloping gently southward, evidently once a portion of Pahute mesa, but now separated from it as a result of erosion. The hills are bare of timber. A small seep, dry in summer, is located in the northern part of the ridge.

GENERAL GEOLOGY.

The formations of these hills, from the oldest to the youngest, are the following: Eureka quartzite, rhyolite, biotite andesite, and basalt.

SEDIMENTARY ROCKS.

Eureka (?) quartzite.—Mr. T. C. Spaulding states that a small hill on the north of the range is composed of lilac-gray quartzite. The quartzite is rather coarse grained and contains some tiny pebbles of kaolinized feldspar and red jasperoid. The similarity of this rock to that of Quartzite Mountain in the Kawich Range indicates that it is probably the Eureka quartzite.

IGNEOUS ROCKS.

Rhyolite.—The rhyolites are similar to those of the Cactus Range and are presumably contemporaneous. Slightly vesicular facies are, however, present.

Biotite andesite.—Masses of dark-gray biotite andesite identical with that of the Cactus Range intrude the rhyolite in the northwestern portion of the hills.

Basalt.—The greater portion of these hills is composed of black basalt with dense groundmass. The phenocrysts are striated feldspar, black pyroxene, and pellucid olivine, the feldspar and pyroxene showing a rude flow alignment. Vertical columnar jointing is well developed in much of the rock. Some tuffaceous sediments with perfect feldspar crystals lie below the basalt on the west side of the area, and it is not improbable that the younger tuffs, as well as the later rhyolite of Goldfield, are here associated with the basalt. This rock is probably of late Pliocene or early Pleistocene age.

MONITOR HILLS.

The Monitor Hills are situated in the north-central part of the area surveyed They are of gentle grade and rise 1,000 feet above the surrounding alluvial slopes.

GENERAL GEOLOGY.

The rocks of the Monitor Hills, the oldest being named first, are granite, Siebert lake beds, rhyolite, and basalt.

SEDIMENTARY ROCKS.

Siebert lake beds.—These hills are in greater portion formed of sandy slopes, from the disintegration of well-bedded tuffaceous sandstones. These are light gray or white in color and usually rather fine-grained, although some facies are conglomeratic. The rounded pebbles are of light-grayish glassy rhyolite. The tuffaceous sandstones show an apparent exposure of 800 feet. These beds are lithologically similar to and are here correlated with the Siebert lake beds of Tonopah, situated 16 miles north of west of the Monitor Hills. These Spurr[a] considers of Miocene age.

IGNEOUS ROCKS.

Granite.—Angular fragments of a medium-grained siliceous granite occur in the basalt described below, indicating that the basalt in its ascent to the surface passed through a mass of granite.

Rhyolite.—Two isolated hillocks to the south of the main hills appear from a distance to be an igneous rock that is lighter in color than basalt, and that is probably rhyolite.

Basalt.—A number of basalt areas are scattered over the Monitor Hills, and this rock caps the summit of the group. The compact facies has a dense blue-black groundmass which is locally spotted with gray blotches one-fourth inch in diameter. The phenocrysts include light-gray glassy feldspar from one-sixteenth to one-half inch in length. Columns of a dark pyroxene and rounded grains of a glassy black substance, presumably olivine, occur. Under the microscope this rock proves to be a holocrystalline olivine basalt. The black color of the olivine in hand specimens is due to its partial alter-

---

[a] Spurr, J. E., Prof. Paper U. S. Geol. Survey No. 42, 1905, p. 69.

Bull. 308—07 m——7

ation to a blood-red serpentine. Vesicular facies of the basalt are usually reddish brown and scoriaceous facies are brick red. In some beds angular fragments of basalt are embedded in a matrix of similar basalt, forming a flow breccia and showing that motion continued in the lava after some portions had solidified. The more dense facies show spheroidal weathering well developed. Vesicular basalt bounds a central sheet of dense basalt and a ropy surface. The original flow surface is seen at places; this indicates that the larger portion of the basalt is without doubt a flow. The largest mass, however, is prob- ably the site of the vent from which the flows occurred.

Where the contact between the basalt and the Siebert lake beds is exposed the basalt is seen to lie upon the eroded surface of the sand- stone. While it is possible that some masses of basalt on the lower hill slopes are flows contemporaneous with the Siebert lake beds or are later sheets, they are more probably portions of the flow which owe their position partly to the uneven surface of the tuffs and partly to later deformation. The basalt, on structural and lithologic grounds, is believed to be contemporaneous with that of Tonopah, which Spurr[a] believes to be of late Miocene or early Pliocene age.

STRUCTURE.

The Siebert lake beds and the basalt flows are in a broad way hori- zontal, although they have been tilted slightly to the northeast. The relative distribution of the basalt and the lake beds is such as to sug- gest that they are cut into a number of blocks by intersecting systems of normal faults.

CACTUS FLAT.

Cactus Flat lies between the Cactus and Kawich ranges. Its center is occupied by a series of playas with north-south trend, which during very exceptional rains merge into one another. They are from 5,330 to 5,350 feet above sea level, and are surrounded by low sand dunes. In this valley the Recent detrital deposits may be thick. The valley from the Cactus to the Kawich Range across Cactus and Gold flats illustrates the steeper slope of valleys near large and high mountain masses and the gentler grade near small and low hills.

GOLD FLAT.

Gold Flat lies between Pahute Mesa and the Cactus and Kawich ranges. Its lowest point, 5,035 feet above sea level, is occupied by a playa, 4 miles west of which, in a small inclosed basin, is another playa. North of the larger playa, at an elevation of 20 to 100 feet above its surface, are a number of round intensely dissected hillocks.

---

[a] Spurr, J. E., Prof. Paper U. S. Geol. Survey No. 42, 1905, p. 69.

These hills show a total thickness of 100 feet of white, well-stratified clay, to be correlated with similar playa deposits of the older alluvium in the Stonewall Flat. Presumably this playa deposit and its associated alluvial slope deposits underlie the Recent gravels of Gold Flat at no considerable depth.

KAWICH RANGE.

TOPOGRAPHY AND GEOGRAPHY.

The Kawich Range was named after Chief Kawich, an Indian who lived at the spring at the head of Breen Creek. The range is a southward continuation of the Hot Creek Range, from which it is separated by a narrow transverse valley.[a] From this gap it extends 50 miles southward to a point where it sinks beneath the lava flows of Pahute Mesa. The range has a well-developed crest line with north-south trend. It is deeply indented by valleys at Rose Spring and at Kawich, while 8 miles north of Kawich the Recent alluvial deposits completely arch the range.

In the rugged northern part of the range the sinuous crest line is in many places a knife-edge flanked by high cliffs; south of Rose Spring the topography is much less rugged. From the main crest line subordinate divides of similar character extend at right angles. The crest line at the north edge of the area mapped has an elevation of 9,000 feet, while in the south end of the range it rarely rises above 7,000 feet. Kawich Peak, with an elevation of 9,500 feet, is the highest summit. Quartzite Mountain, a beautiful symmetrical dome southwest of the village of Kawich, is 7,900 feet high.

The Kawich Range is intensely dissected by many stream channels. The drainage line of Little Mill Canyon, which alone is abnormal, courses east 2 miles, bends sharply to the north, and 1 mile below again turns at right angles and resumes its easterly course to the Reveille Valley. A low divide separates the first segment of the stream from a prominent east-west valley in alignment with it south of the mountain southwest of the town of Eden, and it is probable that Little Mill Creek has captured the upper part of its valley.

On the crest line, near the northern boundary of the region shown in the map, are areas of an older and more mature mountain topography, with gentle slopes, in strong contrast with the rugged character of the surrounding ridges and cliffs. Exposures of rock are few and nowhere conspicuous; the surface is covered with soil and supports some grass. This old topography is in advanced maturity or in old age, but the hills existing on its rolling surface are too numerous to call it a peneplain. Remnants of a similar surface are preserved in the Belted, Amargosa, and Panamint ranges, and in the Panamint

---

[a] Spurr, J. E., Bull. U. S. Geol. Survey No. 208, 1903, p. 181.

the surface was developed prior to the outflow of the Pliocene-Pleistocene (?) basalt. This less accentuated surface, which probably extended completely over the area surveyed, is perhaps of early Pliocene age.

The Kawich Range is rather heavily timbered at its north end above an elevation of 6,500 feet. The timber line becomes higher to the south and neither piñon nor juniper grows south of the detritus-covered gap 8 miles north of the village of Kawich. In consequence of its height and of its rather heavy growth of timber, the Kawich is the best watered range within the area surveyed. The streams of this range have already been described (p. 18). The springs south of the timbered areas are small and some become dry during the summer. The following are rough estimates of the flow of the larger springs in gallons per day: Longstreets Ranch Spring, 3,000 gallons; Stinking Spring, 2,000 gallons; Heenan Water, 1,500 to 2,000 gallons; Rose (sometimes called Wild Horse) Spring, 2,000 gallons; Sumner Spring, 1,500 gallons; Corral Spring, 1,500 gallons; Jarboe Spring, 500 gallons. At Silverbow and Stephanite wells have been sunk in stream gravels and water encountered at depths of 10 to 45 feet.

GENERAL GEOLOGY.

The formations of the Kawich Range, in ascending order, are: Pogonip limestone, Eureka limestone, Lone Mountain limestone, granite, diorite, monzonite porphyry, earlier rhyolite with minor contemporaneous latite and dacite and basalt flows, biotite andesite, dacite, Siebert lake beds (?), later rhyolite, and later basalt and associated andesites.

SEDIMENTARY ROCKS.

Pogonip limestone and Eureka quartzite.—Quartzite Mountain is composed of Paleozoic rocks and from it an arm extends north on the west side of the range and another south on the east side of the range. Small inliers of Paleozoic rocks also occur 2 miles east of Rose Spring, 6 miles west of north of the same point, and 3 miles north of west of Silverbow. Angular fragments of Paleozoic jasperoid, quartzite, and schistose shale are widely distributed in rhyolite, and the vent through which the rhyolites were extruded passed through Paleozoic rocks. Inclusions of Paleozoic rocks also occur in the andesites southeast of Quartzite Mountain.

The sedimentary rocks exposed on Quartzite Mountain consist of 1,200 to 1,500 feet of quartzite underlain by 400 to 600 feet of interbedded quartzites, slaty shales, and limestones. The quartzites of Quartzite Mountain are typically light in color, usually being either white, pinkish white, or gray, although some of the more feldspathic varieties are greenish on account of the development of secondary sericite. Rarely thin laminae of dark gray or black alternate with the white bands. Pure quartzose facies are more abundant than feldspathic, although well-characterized arkoses also occur. While these rocks are largely quartzites some beds are better defined as indurated sandstones. The rock is typically fine to medium grained, but conglomerates with well-rounded pebbles one-half inch in diameter also occur. The pebbles include smoky or pinkish quartz, probably of granitic origin, pink feldspar, compact red jaspilite, and fine-grained biotite schist with phenocrystic muscovite crystals one-eighth inch long. These pebbles can not be certainly referred to any Paleozoic rocks seen in the area surveyed, although they may possibly be derived in part from the Cambrian or from a pre-Cambrian series. The quartzite is medium to heavily bedded. Cross-bedding is common and ripple marks are locally well developed, while sun cracks are rare and confined to the finer grained arkoses. The quartzite is well cemented and both pebbles and matrix are cut by numerous quartz veinlets. The presence of two intersecting systems of quartz veins at an angle of 45° to the bedding planes, one set short and rather wide in the plane of stretching, the other thin, more numerous, and of greater length in the plane of compression, indicates that the quartzite beds have suffered considerable shearing. Muscovite films have been developed on some parting planes. Minor faults occur in the quartzite mass and in some cases cut the quartz veins. Joints are well developed and in consequence the rock fragments are angular blocks.

The rugged inlier of quartzite 3 miles north of east of Silverbow is surrounded by rhyolite. It is similar lithologically to the purer white or pinkish-white quartzites of Quartzite Mountain. Thin quartz veins with vugs and a few seams of pyrite cut it. The quartzite area 6 miles west of north of Rose Spring exposes 400 feet of similar quartzite. Conglomeratic bands with white quartz pebbles one-half inch in diameter are interbedded with the fine-grained bands.

The slaty shale of Quartzite Mountain grades on the one hand into argillaceous sandstone and on the other into clayey limestone. The shale is fine grained and olive green or dark gray in color, although in rare instances bright-green facies occur through the development of chlorite. Secondary mica films are common on the shale parting planes.

The limestone of the Quartzite Mountain section is a siliceous, compact, well-bedded rock of medium grain. In color it is yellow or reddish yellow. Some beds are oolitic, with white calcite cement which over areas one-half inch in diameter is of common orientation and reflects light like a single cleavage face. The texture is the poikilitic of igneous rocks. Suture joints occur, but are not characteristic. In the small area of sedimentary rocks 2 miles east of

Rose Spring finely banded, dark-colored silicified limestone similar to that of the Southern Klondike hills outcrops, and this, while probably Ordovician, may possibly be a Cambrian inlier.

The rocks of Quartzite Mountain appear to have been laid down in a shallow sea. At some periods during the deposition of the lower portion the sea received fragmental material, both fine and medium grained; at others it received but little, in consequence permitting the deposition of limestone. The presence of sun cracks indicates that at times the deposits were even above sea level. The main quartzite mass was deposited in a shallow sea to which medium to coarse fragmental material was rather constantly supplied.

The rocks beneath the quartzite of Quartzite Mountain resemble the interbedded quartzites, slaty shales, and limestones which overlie the Pogonip limestone of the Belted Range. In consequence the lower rocks of Quartzite Mountain are probably the transitional beds of the Pogonip limestone, while the quartzite, notwithstanding its unusual thickness, is the Eureka quartzite.

Lone Mountain limestone.—A small limestone hill protrudes from the Recent detrital deposits $7\frac{1}{2}$ miles south of east of Rose Spring. Mr. T. C. Spaulding collected a specimen which proves to be a gray crystalline limestone with numerous calcite veins. If the structure of the Paleozoic rocks extending northward from Quartzite Mountain beneath the rhyolite of the Kawich Range continues its easterly dip, this limestone lies above the Eureka quartzite and is probably the Lone Mountain limestone of the Eureka section, described by Hague.

Siebert lake beds (?).—Lying unconformably upon the earlier rhyolite and in one case apparently unconformably below the later rhyolite are a number of small exposures of rhyolitic sandstone, usually conglomeratic. These slightly consolidated sandstones are white, gray, or greenish in color and are distinctly bedded. The well-rounded pebbles and bowlders, the larger of which are 1 foot in diameter, include Paleozoic quartzites and jasperoids, granites, diorite porphyry, monzonite porphyry, and the earlier rhyolite. Quartz, feldspar, and biotite crystals are characteristic, and either explosive eruptions of rhyolite occurred during the deposition of these sandstones or the crystals were derived from rhyolites with incoherent groundmasses. The more easterly of the two masses mapped 3 miles north of Kawich is a bedded conglomerate cemented by chalcedony. The conglomerate is better assorted than that of the rhyolitic sandstone and resembles the well-rounded conglomerates of the Amargosa Range. It is certainly younger than the earlier rhyolite, since it carries rhyolite bowlders and is perhaps contemporaneous with the Siebert lake beds.

Spurr[a] found similar rhyolitic sandstones 600 feet thick at the

_____
[a] Spurr, J. E., Bull. U. S. Geol. Survey No. 208, 1903, p. 163.

north end of the Reveille Range. If the correlation of the earlier rhyolite with the Tonopah rhyolite-dacite is correct, these beds are probably the Siebert lake beds, of Miocene age. The last-mentioned exposure may, like the conglomerates of the Grapevine Range, be a shore deposit of this lake.

Post-Jurassic granite.—One mile west of Sumner Spring is an outcrop ·of gray granite porphyry with a dense groundmass containing many phenocrysts of feldspar three-fourths of an inch in length and rounded quartz individuals one-fourth inch in diameter. Under the microscope the groundmass shows as a fine mosaic of quartz and orthoclase with a little plagioclase. With the phenocrysts of orthoclase and embayed quartz are some of plagioclase and very small tablets of biotite altered to muscovite and a rutilelike substance. Apatite, zircon, and magnetite are accessory minerals. The relations of this mass to the earlier rhyolite are poorly exposed, but on lithologic grounds it is considered a post-Jurassic granite porphyry.

Granite inclusions in rhyolite were observed at five widely separated localities. The varieties include fine- or medium-grained biotite granites, coarse-grained muscovite granite, pegmatitic quartz, and biotite granite with porphyritic feldspar crystals similar to the granite of Lone Mountain. Large bowlders of medium-grained biotite granite are so abundant in the Siebert lake beds 4 miles east of north of Silverbow that a granite outcrop was probably near at hand when the sandstone was deposited.

Diorite porphyry.—Bowlders of diorite porphyry are associated with those of granite in the area of Siebert lake beds just mentioned. The rock has a white, finely crystalline groundmass, in which are embedded numerous white striated feldspar phenocrysts with a maximum length of one-fourth inch and smaller biotite phenocrysts. The rock is similar to the diorite porphyry of the Lone Mountain foothills and is probably a later manifestation of the post-Jurassic granite intrusion.

Monzonite porphyry.—Monzonite porphyry occupies an area 4 miles long and 3 miles wide west of Kawich. It has a fine-grained dark or medium-gray groundmass, in which are embedded abundant phenocrysts, many one-fourth inch, a few one-half inch long, of white or greenish-white striated feldspar, brown biotite, and altered hornblende. While both hornblende and biotite are usually present the relative porportion of the two varies widely throughout the area. Oxidation of the iron of the rock produces pink and purple varieties, while kaolinization gives a white color.

Microscopic examination shows the groundmass to be a medium-grained interlocking mosaic of orthoclase and plagioclase. Plagioclase phenocrysts, which are more abundant than those of hornblende or biotite, are usually complex crystals showing albite and, less commonly, Carlsbad and pericline twinning. Zonal growth is rather common. The plagioclase is slightly altered to calcite, zoisite, sericite, and chlorite. Biotite is altered to chlorite, calcite, magnetite, titanite, and epidote. Areas of chlorite, epidote, calcite, and quartz rudely preserve hornblende forms, while a little fresh brown hornblende is present.

Dikes of monzonite porphyry cut the Paleozoic sediments and the main mass is probably an eroded stock. The outcrops of monzonite porphyry are cut by rather closely spaced joints and in consequence break down into angular blocks. Spheroidal weathering is here and there poorly developed.

The monzonite porphyry in places appears to lie upon the eroded edges of the Paleozoic sedimentary rocks and inclusions of monzonite porphyry occur in the rhyolite on its northern border. This rock is rather similar to the biotite-hornblende latite of the Cactus Range and more closely resembles the quartz-monzonite porphyry of Stonewall, Shoshone, and Skull mountains. These rocks are in a broad way contemporaneous and of late Eocene age.

Earlier rhyolite.—Earlier rhyolite with associated latite and dacite forms most of the mountain range north of an east-west line 2 miles north of Kawich. Small areas occur south of this line and in the Reveille Valley. The rhyolite south of Rose Spring and on the flanks of the range north of this point is unquestionably a flow. The crest of the range north of Rose Spring is composed of compact rhyolite, in which flow bedding is absent or very obscure. It is probable that the crest in part or as a whole lies upon the fissure from which the rock was extruded. Some of the masses of rhyolite in the vicinity of Kawich are perhaps dikes intruded in the monzonite porphyry, and the poorly exposed rhyolite in the Paleozoic quartzite $6\frac{1}{2}$ miles west of north of Rose Spring may also be a dike.

These acidic rocks include many facies. Perhaps the predominant type is a reddish rhyolite with compact flinty groundmass, containing abundant medium-sized phenocrysts of glassy orthoclase and slightly smoky quartz, with fewer and smaller phenocrysts of black mica, which in some instances are absent. The quartz phenocrysts are, as a rule, rounded crystals and only a few show the dihexagonal pyramid and prism. Other rhyolites are gray or brown in color, while some black glasses with the normal phenocrysts also occur. In the more glassy types flow lines, spherulites, and perlitic parting are common. The microscope shows the groundmass of the rhyolite to be of turbid brown glass, in which flow lines and spherulites are well developed locally. The phenocrysts are in some instances well formed and in others are angular fragments, showing that flow continued while the lava was very viscous. Biotite is fairly abundant in some slides, while in others it is represented by ill-defined pseudomorphs.

Quartz latites are widely interbedded with the rhyolites, being particularly well exposed at Stinking Spring. They are characterized in the hand specimen by the presence of phenocrysts of both striated and unstriated feldspar and a hornblende-like mineral which reaches a maximum length of one-fourth inch. Microscopic examination shows a glassy groundmass containing as phenocrysts quartz, orthoclase, plagioclase, and hornblende or augite. Both feldspars are often zonally built. Magnetite and apatite are common accessories.

Brown glassy dacites are less widely distributed. Microscopic examination shows a rock similar to the quartz latite, except that plagioclase is present to the almost total exclusion of orthoclase, and ilmenite accompanies magnetite.

Many beds of the acidic series are flow breccias, including angular fragments of texturally similar or dissimilar acidic rocks, indicating that portions of the flow solidified prior to the cessation of movement in other parts. White, slightly porous pieces of rhyolitic glass, elongated in the direction of flow, are common in certain areas. Similar inclusions characterize the earlier rhyolite of Goldfield and the rhyolite of Bullfrog.

About 1½ miles east of Rose Spring is an area of consolidated white shaly sandstones and coarse, well-bedded sandstones. The microscope shows that the rock is a typical tuffaceous sandstone of rhyolitic composition. It was probably deposited in a small body of water contemporaneous with the outflow of the earlier rhyolite.

The acidic series imparts to the Kawich Range a white, yellow, or red color, and in places all these are distributed in broad bands either horizontally or slightly tilted. Spires are developed on canyon walls in jointed rhyolite, the same rock forming rounded bosses in the broader, more mature valleys. South of the detrital gap in the range the hills have a marked northwest-southeast alignment parallel to the strike of resistant rhyolite flow bands.

The earlier rhyolite of the Kawich Range lies below supposed Siebert lake beds and is similar to and probably contemporaneous with the rhyolite of the Belted, Cactus, and Reveille ranges and the Bullfrog Hills. In a general way the series is comparable with the pre-Siebert rhyolites and dacites of the Tonopah district,[a] which are probably of early Miocene age.

Earlier basalt.—Minor flows of basalt were extruded contemporaneously with the earlier rhyolite. The small mass of dense black basalt with sparse, medium-sized, glassy feldspar phenocrysts three-fourths of a mile east of Silverbow is probably such flow. This rock on microscopic examination proves to be a holocrystalline

---

a Spurr, J. E., Prof. Paper U. S. Geol. Survey No. 42, 1905, pp. 37–43.

olivine- and augite-bearing basalt, somewhat altered. Orthoclase phenocrysts are also present. Small inclusions of vesicular basalt occur in rhyolite 1½ miles north of Silverbow. Basaltic flows contemporaneous with rhyolite also occur in the Amargosa Range and the Bullfrog Hills.

Biotite andesite.—One mile northwest of Silverbow a small hill surrounded by alluvium is composed of a fine-grained dense andesite of gray color. The small phenocrysts, which are in rude flow orientation, consist of white striated feldspars and more numerous smaller biotites. Southeast of Blakes Camp is a poorly exposed mass of a greenish-gray andesitic rock, considerably altered. Biotite and striated feldspars one-eighth inch long are equally abundant. Under the microscope the glassy groundmass is seen to be subordinate to the phenocrysts, which consist of medium-sized simple crystals of plagioclase, biotite, and a few deeply embayed orthoclase individuals. The plagioclase is considerably altered to calcite, and the sericite and the biotite to chlorite. It is probable that these two exposures are to be correlated with one another and with the biotite-andesite intrusive in the rhyolite of the Cactus Range.

Dacite.—In the vicinity of Rose Spring are five small areas of dacite. Another small mass lies 1 mile east of Stinking Spring and a rather similar rock is exposed 3 miles south of Georges Water. The dacite has a compact aphanitic groundmass of medium to dark-gray or brown color. The phenocrysts equal the groundmass in bulk, and consist of medium-sized glassy striated feldspars, hornblende, mica, and quartz. The quartz, while not abundant, is constantly present. A few well-formed crystals of striated feldspar, evidently of a later origin than those just mentioned, are from three-fourths of an inch to 1½ inches long. These are white or gray in color and include tiny biotite plates and hornblende columns. Weathering brings out clearly the zonal structure, while Carlsbad twins are sometimes seen. The phenocrysts near contacts with older rocks possess a parallel flow orientation. Microscopic examination shows the rock to have a felty groundmass of plagioclase laths with a few orthoclase and quartz grains. Many of the smaller plagioclase phenocrysts, which are usually compound and slightly rounded by magmatic corrosion, show beautiful zonal growth. Plagioclase is ordinarily fresh, although calcite is locally developed from it. One or more quartz phenocrysts deeply embayed are present in most of the thin sections. Biotite, usually surrounded by a magnetite reaction rim, is abundant. Forms characteristic of hornblende and heavily rimmed by magnetite grains are completely altered, calcite, in places accompanied by serpentine, and magnetite being the alteration products. Magnetite and apatite are accessory minerals. The dacite occurs in dikes and irregular intrusive masses which cut the earlier rhyolite 2 miles southwest of Sumner Spring. The mass 1½ miles east of Rose Spring is intrusive in shaly sediments that are probably interbedded with the earlier rhyolite. This dacite is lithologically identical with the dacite of the Goldfield hills and the two are probably contemporaneous and of middle Miocene age.

Later rhyolite.—Forming the gently sloping flanks of the range west of Silverbow and capping some of the hills to the south of the same village is a brown, dense, and in many places glassy rhyolite characteristically poor in biotite and rich in quartz phenocrysts. Vesicles, elongated parallel to the flow banding, and vertical joints of cooling occur. In some places this rhyolite apparently lies upon the slightly eroded surface of the earlier rhyolite, and 5 miles south of Blakes Camp it appears, at a distance, to overlie the Siebert lake beds. The later rock has suffered much less erosion than the earlier rhyolite. The rock, so far as seen, was not altered by the mineralizing waters which formed the ore deposits. This rhyolite is younger than the supposed Siebert lake beds and is perhaps about contemporaneous with the later rhyolite of the Belted Range.

Later basalt and associated andesites.—Basalt caps the Kawich Range south of Quartzite Mountain. A number of basalt hills outcrop in the Reveille Valley, and on the east side of the range are a number of low buttes partly of the same rock. The basalt of the south end of the range is black, dense, and more or less vesicular. The cone on the east flank of the mountains 10 miles east of north of Kawich is composed of a similar rock, in which a few small feldspar laths are the only conspicuous phenocrysts. Olivine basalts occur in the Reveille Valley. Basalt also forms a mesa 5 miles west of Silverbow. Andesite is associated with some of the basalt. The andesitic rocks of the Reveille Valley have a dense mottled reddish-gray or lilac-gray groundmass, which is locally vesicular. The medium-sized phenocrysts, which equal the base in mass, include glassy striated feldspar laths, many of which are twinned according to the Carlsbad law; fewer biotite plates, and hornblende or pyroxene columns. Spheroidal weathering is rather characteristic. Flow breccias occur in bands, and the rock appears to be closely related to the associated basalts, each partially burying the small earlier rhyolite outcrop. A similar rock, although practically lacking biotite, forms a number of low, flat-topped hills 6 miles south of Kawich. Phenocrysts are abundant in some facies and in others lacking. The rock occurs in flows which dip to the east. The hill 6 miles N. 30° E. of Kawich is capped by vesicular basalt beneath which are gray andesitic rocks with glassy striated feldspar phenocrysts up to one-fourth inch in diameter.

The basalt is similar to that of the Belted and Reveille ranges and is probably of late Pliocene or early Pleistocene age. The andesites are doubtless earlier manifestations of the same periods of volcanism.

The Paleozoic rocks are complexly folded (see fig. 8). The Quartz-ite Mountain section is a north-south monocline with dips of 10° to 30° E. The Paleozoic rocks on the west side of the Belted Range dip to the east, and the Kawich Valley probably overlies a gentle syncline, although the structure may be much more complex. The ver-tical strata of the inlier 3 miles north of east of Silverbow strike east and west. The quartzite 6 miles west of north of Rose Spring strikes N. 75° W. and dips 45° S. Zones of brecciation, indicating that this pre-Tertiary folding was accompanied with some faulting, are not un-usual. The presence in the volcanic rocks of Paleozoic inliers proves that prior to the inauguration of the Tertiary volcanism the Pale-ozoic rocks of the Kawich Range had a diversified topography. The mountain range then may have been as rugged as now, since the val-leys on either side were doubtless at lower levels than at present.

It is unknown whether or not the normal faults in the monzonite porphyry at Kawich, which are in part older and in part younger than the ore deposits, were formed prior to the outflow of the rhyolite.

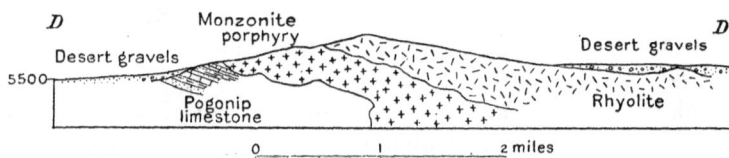

FIG. 8.—East-west section across Kawich Range 2½ miles north of Kawich.

The rhyolite series north of Rose Spring is a monocline with north-south strike and low easterly dip. Minor folds of similar strike are superimposed upon the monocline. At the detrital gap the strike changes to northwest with a dip of 10° to 20° NE. This uplift occurred after the deposition of the Siebert lake beds, since they lie on the very crest of the range. Joints were developed in the rhyolite contemporaneously with the folding. Sheeting parallel to the folds is rather characteristic, although cross joints occur. After the out-flow of the basalt the range was uplifted, as is shown by the fact that the flows southeast of Quartzite Mountain have a decided dip to the east. An undetermined amount of normal faulting accompanied the post-rhyolitic elevation, and in certain areas, notably in the shales associated with the rhyolite 1½ miles east of Rose Spring, the faulting is very complex.

ECONOMIC GEOLOGY.

Mining activity in the Kawich Range centers in three major camps—Silverbow, Kawich, and Eden—and in two minor camps—Blakes Camp and Southern Kawich.

Silverbow is situated in the canyon of Breen Creek, on the west side of Kawich Range, 43 miles north of east of Goldfield. The principal prospects are within 3 miles to the east, northeast, and northwest of the village. The first locations were made in November, 1904. When visited by the writer (July 20, 1905) Silverbow was the supply center for several hundred men, and a number of shallow prospect holes had been sunk.

The ore deposits lie in rhyolite, which in the vicinity of veins is either kaolinized or silicified, the silicification being perhaps more common. (See p. 47.) Either type may be intensely stained red, brown, or yellow by iron oxides.

The more important prospects are located on parallel quartz veins or lenses, which widen, thin, and in places give out, forming mineralized bands whose strike in the district is in many cases north of west. The individual quartz veins vary in width from a fraction of an inch to 5 feet, and many of these are connected by minor cross veins. While in some instances the quartz was deposited along premineral faults, in others it occurs along joints, which locally form intersecting systems. Quartz likewise fills many of the spaces caused by brecciation and forms in solution cavities in the rhyolite. The quartz is, as a rule, white and translucent or colorless and transparent, although in the Blazier tunnel a single vein of amethyst-purple color was noted. A thin section of quartz from the Wheaton and McGonagill prospects is a quartz mosaic of very uneven grain. In the mosaic are zonally grown crystals of quartz, and finely divided silver sulphides dust some of the bands of growth. The borders of these quartz crystals appear to be formed of quartz fibers, radiating from the center. Crystal-lined vugs are common in this vein quartz, and crustification is beautifully developed in many specimens, fortification, agate, and mammillary forms being common.

The quartz is more or less stained by iron salts, rarely by malachite. In the quartz specks of stephanite, ruby silver, silver chloride, and probably other silver ores occur. The silver chloride is at least a secondary mineral and to a limited extent is disseminated in the country rock. Gold occurs free. Silver is the predominant metal, and $1 in gold to $3 in silver is perhaps an average for the whole camp, although in some prospects the silver values are 20 times those of gold. The ore runs from $6 to $250 per ton, though some reports give higher values. Since the writer's visit strikes in which gold predominates over silver have been reported. Rarely values occur in limonite stringers in the country rock, while in several prospects a greasy crushed rhyolite constitutes the ore.

The striking similarity between the ore deposits of Silverbow and certain of those of Goldfield is worthy of note. Ascending water carrying silica and metallic salts in solution appears to have deposited its burden largely in preexisting cavities. Simultaneously waters wandering from the main channels silicified contiguous portions of the rhyolite. Kaolinization is presumably a later phenomenon. Movement has occurred in the veins since silicification, and they have been more or less faulted and brecciated. Surface waters have altered silver sulphides, pyrite, and probably less amounts of chalcopyrite to the corresponding chloride, oxides, and carbonates, while secondary sulphides have also been formed. Gold, probably originally in pyrite, was simultaneously set free.

Timber and water are abundant near the mines. Goldfield is about 50 miles distant by road.

BLAKES CAMP.

Blakes Camp is located on one of the earlier rhyolite inliers which protrude from the " wash " 12 miles northwest of Silverbow. The prospect was discovered late in June, 1905. A fault with well-defined walls coursing N. 80° E. cuts the rhyolite, and in its immediate vicinity the country rock is silicified. Along this fault for a distance of 600 feet a zone from 18 to 24 inches wide is crushed to a fine clay, in which are embedded slickensided fragments. The ground-up rhyolite is white or stained by hematite or limonite and to a less extent by manganese dioxide. This material is said to pan about $15 per ton of free gold. Quartz veins in rhyolite north and east of Blakes Camp are said to carry both gold and silver.

EDEN.

The mining camp of Eden is situated near the mouth of Little Mill Creek Canyon, on the east side of the Kawich Range. The first locations were made February 20, 1905, by John Adams.

Three distinct vertical zones of mineralization cross Little Mill Creek from north-northeast to south-southwest, respectively, one-fourth, one-half, and five-eighths mile above the old mill at the road forks. The country rock is rhyolite, which in many places is intensely silicified parallel to the mineralized zones. While the contact between the silicified and the unsilicified rhyolite is as a rule rather sharp, in some places there is transition. The quartz and silicified rhyolite stand out in distinct walls, aptly called by the miners " dikes," which follow the mineralized zones.

The central zone has been more developed than the others and was studied in more detail. This zone is situated along a line of faulting and brecciation and the rhyolite near the zone is mashed. The mineralized zone varies from a vein of quartz 3 feet wide with a few rhyolite horses to a band of silicified rhyolite 8 feet wide, intensely netted by quartz stringers. Parallel quartz veins more or less continuous and connected by cross veins are transitional between the two forms along the strike. The quartz, much of it crustified, is white and more or less iron stained. Finely divided sulphides impart a blue tinge to some of the quartz, and the parallel position in thin bands of the white and blue quartz shows clearly the contemporaneous origin of the quartz and sulphides. Vugs, usually elongated, parallel to the walls of the veins, are characteristic. Colorless quartz crystals are common in these vugs, while botryoidal quartz walls are less common. The vein in some portions is composed of clusters of quartz plates an inch in length set at right angles to the vein walls. These suggest that the quartz in parts is a replacement of calcite. The normal crystalline quartz grades into a compact white flinty variety, which is said to be a good carrier of gold. Under the microscope this appears as a cryptocrystalline mosaic of quartz with a little sericite. This central zone is a silver bearer, the gold values being one-tenth those of silver. As silver ores, ruby, and native silver are reported intimately associated, secondary silver chloride in knobs and wire-like masses is widely distributed in the vugs, and the silicified rhyolite itself carries values. Iron sulphide was noted in a few small specks in quartz vein.

The upper zone of mineralization is said to be very similar to the central zone. The lower zone is also similar to the central, although there is perhaps less parallelism and constancy of the quartz veins. The strike is more northerly. Pyrite locally impregnates the rhyolite. The flinty quartz already mentioned is abundant in this ledge, and the values are largely gold. The resemblance of the Eden deposits to those of Silverbow is striking, and the genesis was doubtless similar and contemporaneous.

Water and fuel are abundant. Tonopah, the supply depot, is 70 miles distant.

KAWICH.

Kawich, in the Gold Reed mining district, lies in a detrital embayment on the east side of the Kawich Range. The town is 54 miles south of east of Goldfield. The first locations were made in December, 1904, and early in the spring of 1905 several hundred men rushed to the camp. When visited (August 4, 1905) 10 miners were at work. Two shafts have been sunk 150 feet and several thousand feet of drifts have been driven. The mines are situated on a gently sloping area of wash, from which protrude numerous small rugged knobs of monzonite porphyry and smooth domes of rhyolite. The monzonite porphyry has so far been the ore bearer.

The first locations were made on the property of the Gold Reed Mining Company. Here a rugged outcrop of intensely silicified monzonite porphyry, dark brown on weathered surfaces, contains in the limonite-stained casts of its many phenocrysts plates of hackly gold,

some of which are an inch in diameter. This altered rock has a conchoidal fracture. The microscope shows the rock to be composed of
interlocking quartz grains in which there is some finely divided material, possibly of a clayey nature. The areas once occupied by feldspar
phenocrysts contain a slightly higher percentage of this turbid substance, while some sagenitic webs of rutile indicate the former presence of biotite. Associated with the silicified porphyry are white
kaolinized facies which are locally stained red or purplish by iron
oxides. Numerous other areas of silicified porphyry have been located
by prospectors, but at none was gold visible to the eye, although many
panned gold. The boundary between the silicified and unsilicified
porphyry is usually sharp. Strong iron stains and some dendrites of
manganese dioxide are associated with the gold. The iron stains
probably point to pyrite as the original source of the ore, an inference
strengthened by the presence of gypsum. This mineral is rather
closely associated with the ore, in some places filling cavities in it and
in others filling joint and fracture planes in the country rock. The
gypsum occurring in the country rock is fibrous, the fibers being set
at right angles to the walls. The residual soil near ore deposits is
locally cemented by gypsum. Gypsum was probably formed by the
interaction of iron sulphate, set free by the oxidation of pyrite, and
of calcium-bearing solutions derived from the decomposition of the
plagioclase of the monzonite porphyry. Quartz veinlets cut the silicified porphyry, but these are reported to carry no values. Certain
clear phenocryst-like quartz areas indicate that silicification continued after the removal of the phenocrysts usual in monzonite porphyry.
In the deepest shafts the silicified porphyry holds its width, although
up to the time of this examination the values were low except at the
surface. At the Gold Reed mine good values have been obtained in a
kaolinitic substance found in the cavities of the silicified porphyry at
the 100-foot level.

Iron pyrite has been encountered on the 150-foot level at the Gold
Reed mine, disseminated in unsilicified monzonite porphyry. Its
assay value is low. Several thin veins of pyrite occur on the lower
levels of the Diamond No. 2. These veins are banded and vugs occur
here and there in the center. This pyrite is said to assay $35 in gold
per ton.

The silicified monzonite porphyry has been considerably faulted
and slickensides and breccias are common. At the Gold Reed mine
the faulting has been very complex, and wedge-shaped cavities,
which reach a maximum length of 5 feet, appear to be due to faulting. At the Chief Kawich mine the faults strike N. 50° W., parallel
to the silicified-porphyry boundary, and dip 80° SW. At the Diamond No. 2 the faults dip 45° E. The step-like offsets in the silicified porphyry on which the Chief Kawich and the Gold Reed mines

are situated are rather suggestive of faulting, although the effect may be due to a chance distribution of several small masses of silicified porphyry.

The resemblance between the Kawich and Gold Crater deposits is close. It is probable that waters bearing silica in solution arose along joint and fault planes and that silica more or less completely replaced the rock contiguous to the fractures. Pyrite was probably simultaneously deposited, partly in the country rock and partly in open fissures. Since the silicification the veins have been faulted. The kaolinization of certain portions of the porphyry was probably contemporaneous with the surface alteration of the original sulphides. From the lean character of the pyrite so far encountered it is probable that the rich surface gold deposits have been concentrated from the pyrite contained in many feet of porphyry which have been removed by erosion, but it is by no means impossible that rich pockets of gold may be found from time to time near the surface.

Water is obtained at Cliff Spring, 12 miles east of the settlement, and piñon grows on the Kawich Range 8 miles to the north. Tonopah is 70 miles distant by road.

SOUTHERN KAWICH.

In the blue-black Pogonip limestone on the northwest slope of Twin Peaks, 6 miles south of Kawich, prospects are located on a quartz vein 1½ feet thick. The quartz is strongly stained by limonite and is said to carry free gold.

AREAS FAVORABLE TO PROSPECTING.

The areas noted in the progress of work in which occur altered rocks of the same character as those which accompany the ore deposits at the camps already described are shown in fig. 4. Other areas of similarly altered rock doubtless exist, although it is believed that the largest are mapped. Of the rhyolite areas shown those near Silverbow and Eden and east of Blakes Camp contain an unusual number of quartz veins. The monzonite porphyry in which the ore deposits of Kawich occur is confined to the immediate vicinity of the town. The areas of altered monzonite porphyry have already been located and the future of this form of ore deposits depends on the development of these areas.

Quartz veins occur in the Paleozoic sedimentary rocks, particularly in quartzite. So far as seen these veins are narrow and of no great length, and with the exception of the Southern Kawich veins the quartz is apparently barren. Quartz veins, probably barren, on the western side of Quartzite Mountain contain tablets of specular hematite.

Reveille Valley lies between the Kawich and Reveille ranges. The depression is separated into two parts by a low gravel divide north of its center, the northern portion draining northward and the southern portion draining southeastward into Railroad Valley. The Recent gravels are probably nowhere thick in Reveille Valley. Basalt flows underlie much of the area and form hills in the valleys and scarps along the drain line. Fig. 9 is a section across the valley, drawn north of west to the Kawich Range from a point on the Reveille Range 6 miles north of Reveille Peak. A single mass of early Miocene rhyolite is exposed beneath the basalt northeast of Sumner Spring.

The Kawich Valley lies between the Kawich and Belted ranges. It is separated from Railroad Valley to the northeast by a low gravel divide and from Reveille Valley to the north by hills of basalt. Kawich Valley is notable for its even grade, since the mountains on either side have approximately an equal elevation and mass. Its

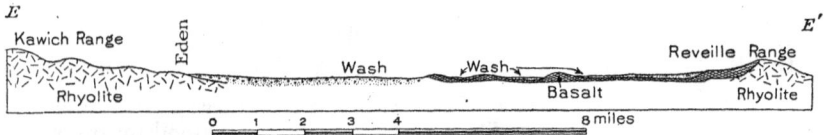

Fig. 9.—Section across north end of Reveille Valley, showing an inclosed valley partially underlain by basalt.

playas lie at an elevation of about 5,800 feet above sea level. A few rounded exposures of older alluvium lie 20 feet above the small playa on the road from Kawich to Indian Spring. These are similar to those of Gold Flat.

TOPOGRAPHY AND GEOGRAPHY.

The Reveille Range is separated from the Pancake Range on the north by a narrow transverse gap and from the Belted Range on the south by Railroad Valley. In former descriptions and in the parlance of prospectors the Reveille and the Belted ranges are merged into one, the name Reveille being applied not only to that range itself but also to the north end of the Belted Range. As here defined, each range is a unit, separated by a transverse valley from 2,500 to 3,000 feet below the crest line of the ranges. These two north-south ranges are, however, aligned, and the structure of the Paleozoic rocks indicates that prior to the long period of erosion which preceded the Tertiary volcanism they were connected.

The Reveille is a narrow, rugged mountain range with a crest line of rather sharp peaks culminating near its south end in Reveille Peak, 8,900 feet high. The eastern slope, descending to the deep Railroad Valley, is the more precipitous. The crest line against which deep, narrow canyons head, is in places a knife-edge flanked on either side by cliffs hundreds of feet high. The more profound of these canyons are 1,000 feet deep and are notable for the forests of rhyolite spires which cover their precipitous sides. The rhyolite is everywhere characterized by a rugged topography. Fang Ridge is a prominent part of the range from which protrude fantastic domes and spires strikingly like the fangs and molar teeth of an ancient monster. The " teeth " are from 200 to 300 feet high. The smooth slopes and shallow, straight-walled canyons of the flanking masses of basalt, which in some cases almost bridge over the range, are in strong contrast with the intensely dissected rhyolite.

The crest line of the range is covered by a fair growth of piñon and juniper, which at the northern border of the area surveyed on the west side descend to the alluvial slopes. Several large springs are located at Old Reveille, 1½ miles north of the boundary of this area, and a pipe line extends from these to New Reveille.

GENERAL GEOLOGY.

The formations of the Reveille Range include, in ascending order, the Weber conglomerate, Pennsylvanian limestone, rhyolite, and basalt.

SEDIMENTARY ROCKS.

Weber conglomerate.—North of the area mapped from Old Reveille to the summit, between that town and New Reveille, the predominant formation is a white massive quartzite, of which from 1,000 to 2,000 feet is exposed. If, as is apparently the case, this quartzite underlies the Pennsylvanian limestone it is the Weber conglomerate, although lithologically it is more nearly related to the Prospect Mountain (Cambrian) or Eureka (Silurian) quartzite.

Pennsylvanian limestone.—Overlying this quartzite, apparently conformably on the west side of the range near Reveille, is 2,000 feet or more of dark-gray to black compact, massively bedded limestone. With it are associated bands of black flint and thin beds of argillaceous limestone. The limestone is cut in every direction by veinlets of calcite. The formation is similar to the Pennsylvanian limestone of the Belted Range, and poorly preserved fossils from a bowlder near New Reveille are considered by Dr. George H. Girty to be probably of Pennsylvanian age. Inclusions of flint are common in the rhyolite, described in the following paragraphs, near its contact with the Paleozoic rocks.

Rhyolite.—The major portion of the Reveille Range is composed of rhyolite, which has a flinty, glassy groundmass, flow banding being common. Gray and pink are the predominant colors, although red, black, and brown are not unusual. Phenocrysts are abundant and equal the base in mass. Colorless or smoky quartz and glassy feldspar phenocrysts of medium size predominate over the smaller individuals of biotite. Weathered surfaces show feldspar whitened or represented by casts, biotite flush with the surface, and quartz in relief. Thin sections under the microscope present a brown, glassy groundmass, exhibiting well-defined flow flexures, and over small areas a spherulitic structure. A small portion of the feldspar is an acidic plagioclase, while many of the biotite phenocrysts were bent prior to the solidification of the lava. Magnetite, zircon, and apatite are also present.

The rhyolite is a series of flows, and in consequence over large areas it weathers like a well-bedded sedimentary rock, a resemblance magnified by the presence of a north-south sheeting. Vertical joints are locally present, and where flow banding is not prominent the rhyolite weathers into spires from 30 to 150 feet high and from 5 to 25 feet in diameter. In places these are single and spring from a common base. Where no parting planes are prominent the rhyolite displays the well-rounded domes and bosses characteristic of a granite country.

The rhyolite of the Reveille Range is similar lithologically to that of the Kawich Range and has suffered equal deformation and erosion. The two are doubtless contemporaneous in age and are probably early Miocene.

Basalt.—Lying upon the rhyolite are basalt flows which usually occur flanking the range, but in one case extend from the eastern base across the crest line. The flows were once much more extensive and perhaps completely covered the range 4 miles north of Reveille Peak. Basalt also forms isolated hills in the alluvial deposits. As shown in fig. 9, these are probably connected with flows on the flanks of the range. The basalt varies from a dense black rock to a vesicular slag of red color. The vesicles reach a maximum diameter of 1 inch and are elongated in the plane of flow. The phenocrysts are subordinate in bulk to the groundmass, and include conspicuous gray or white striated feldspar laths up to three-fourths of an inch long, rather obscure yellow altered olivines, and black pyroxene columns. Weathered basalt breaks down into spheroidal masses surrounded by thin concentric shells.

On microscopic examination the rock proves to be an augite-olivine basalt with glassy base. Abundant plagioclase laths, fewer augite crystals, rounded olivines, and magnetites are the phenocrysts.

The basalt is similar lithologically to the later basalt and is probably of Pliocene age, since the amount of erosion suffered renders its effusion in Pleistocene time improbable.

STRUCTURE.

The Paleozoic rocks are folded into a monocline striking north and south and dipping 15° to 30° W., but a corresponding eastward dipping arm of an anticline may be covered by the rhyolite of the eastern side of the range. Minor faults are present. Prior to the effusion of the rhyolite the rocks formed a rugged mountain range. (See fig. 10.) After the outflow of the rhyolite the range was subjected to orogenic movements, the compressive force acting from the east or the west, and the mountains were again uplifted, the rhyolite was slightly crumpled, and a north-south sheeting was produced. When the basalt flowed out the range was rugged and probably approximately as high as at present. The small peak of rhyolite surrounded by basalt 6 miles west of north of Reveille Peak was probably never completely covered by basalt. A decided thickening of the basalt occurs at the base of the mountains, but the dip of the flow

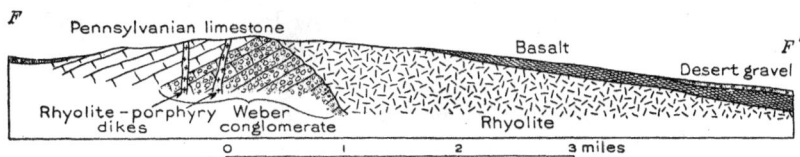

FIG. 10.—East-west section across Reveille Range at northern boundary of area surveyed.

surface, in places 15° to 18°, is probably in part due to recent uplifts of the range. The vent from which the basalt was extruded was probably near the crest of the range 7 miles north of Reveille Peak.

ECONOMIC GEOLOGY.

The abandoned mining camp of Old Reveille lies 1½ miles north of the area surveyed. The Gila mine, one of the best known of those extensively operated thirty years ago, is situated in Paleozoic limestones and quartzites near the rhyolite contact. The ore occurs in quartz veins and stringers ramifying through brecciated zones in the Paleozoic rocks. The ores on the dump include malachite, azurite, cerussite, and galena, with quartz and gypsum as gangues. Malachite here and there coats the surface of the dump and has doubtless been deposited since the mines were abandoned. Recently considerable work has been done on similar deposits at New Reveille.[a] The rhyolite of the Reveille Range, where examined, has not suffered the alterations which usually accompany ore deposits.

---

[a] During the field work, New Reveille was believed by the topographers to be outside the boundary of the area mapped, but more accurate computations have placed it within. In the pressure of reconnaissance work the mines were not examined.

Railroad Valley is a long depression between the Pancake and Reveille ranges on the west and the Grant and Quinn Canyon ranges on the east. Its south end lies in the northeast corner of the area mapped and is of interest because it has two well-developed north-south drainage lines. The playa is 5,106 feet above sea level.

BELTED RANGE.

TOPOGRAPHY AND GEOGRAPHY.

The north end of the mountain range east of Kawich Valley is called by prospectors the Reveille Range, while the south end is known as the Belted Range. The undesirability of using two names for a single mountain range is patent, and as the range immediately north has long been called the Reveille Range the name Belted will here be used for the whole range east of Kawich Valley. Thus defined, the Belted Range extends from the junction of Reveille, Kawich, and Railroad valleys southward 49 miles to Shoshone Mountain, which trends east and west. From the north end of the range the crest courses southward to Wheelbarrow Peak, where it turns to the south-southwest. The range in the latitude of Oak Spring is 13 miles wide. The crest line has an average elevation of 7,000 to 8,000 feet and lies in a median position, except in the central portion of the range, where the long eastward-sloping basalt mesas throw it nearer the western base. Wheelbarrow Peak and a dome 3 miles north of it reach an elevation of 8,600 feet; Belted Peak is 8,340 feet high.

A gentle valley separates the Pogonip limestone from the rhyolite at the north end of the range and a parallel detrital embayment lies west of Cliff Spring. The Pogonip limestone forms hills and ridges elongated parallel to their strike. The Pennsylvanian limestone where metamorphosed by granite forms rugged ridges, but elsewhere its topography is much less accentuated, the ridges being comparatively low and smooth. Minor dip-slope elements are present in all the limestone formations, as well as in the earlier rhyolite near Cliff Spring. The Weber conglomerate forms broad domelike ridges of gentle slope. The earlier rhyolite forms domes and cone-shaped hills with which cliffs are combined near the crest line. Vertical joints are less prominent in the earlier rhyolite of the Belted Range than in that of the Reveille Range, and in consequence spires are uncommon except to the north of Wheelbarrow Peak. The later rhyolite caps prominent buttes, like Oak Spring Butte, whose sides are formed by the soft Siebert lake beds. The basalt mesa country is characterized by level slopes deeply cut by straight-walled canyons.

Remnants of a more mature mountain surface surrounded by recent cliffs and rugged hills occur on Wheelbarrow Peak, on the high peak 3 miles north of it, and in the rhyolite area 5 miles west of south of Wheelbarrow Peak. The old land surface extends 2 miles east of the first-named points. The smooth and gentle slope of this surface is shown on the map, although it is somewhat obscured by the use of 100-foot contours on so small a scale as 1 inch to 4 miles. These mature areas are characterized by few and inconspicuous rock outcrops and a deep soil covered by grass and trees. The contrast between this topography and the sharp ridges and cliffs, practically all a single rock exposure, of the younger topography is very great. Since this old mountain surface developed the Belted Range has been tilted to the east and the older surface has been completely removed from the west versant of the range. (See fig. 11.)

The range north of Belted Peak is bare, but south of this point there is a fair growth of piñon and juniper above 6,500 feet, although the southern quartzite and slate hills are as a rule without timber. A few oak shrubs grow in moist places. The tree yucca and the Spanish bayonet are common on the upper alluvial slopes in the vicinity

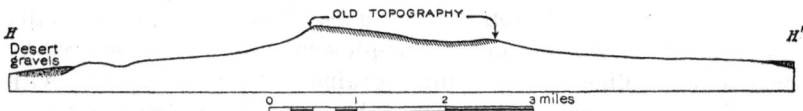

FIG. 11.—East-west profile across Belted Range 3 miles north of Wheelbarrow Peak, showing mature character and present eastward tilt of Pliocene surface.

of Oak Spring. Cliff and Indian springs each flow about 500 gallons of water per day; Wheelbarrow Spring is weak and is probably dry during a part of the summer. Oak, Whiterock, and Captain Jack springs all flow from the Siebert lake beds. Oak and Whiterock springs each flow from 1,500 to 3,000 gallons a day; Captain Jack Spring is considerably smaller.

GENERAL GEOLOGY.

The formations of the Belted Range, from the oldest to the youngest, are as follows: Pogonip limestone, Weber conglomerate, Pennsylvanian limestone, monzonite, post-Jurassic granite, andesite, earlier rhyolite, Siebert lake beds, later rhyolite, and basalt.

SEDIMENTARY ROCKS.

Pogonip limestone.—The Pogonip limestone forms low hogback hills on the western border of the Belted Range from a point 1 mile north of the Kawich-Indian Spring road to the north end of the range. Inclusions of the same rocks occur in the rhyolite and are particularly abundant near its contact with the limestone. A thickness of 4,000 feet of stratified rocks is exposed from a point 2 miles southwest of Belted Peak westward to the detrital deposits mantling the range. The section contains no important stratigraphic break, and while shales and quartzites are interbedded, limestone is the predominant rock. The limestone is typically fine grained, crystalline, and dark gray or black in color. The bedding planes are as a rule massive. Thinner beds of a yellow or brown, more coarsely crystalline limestone are also present, and a single stratum of oolitic limestone was observed. Intraformational conglomerates and cross-bedding, each an indication of shallow-water deposition, occur throughout the section. Nodules and fewer lenses of black chert are present in places. The nodules are in part original, but since the bedding planes in certain instances cut them, they are also in part secondary. A related phenomenon is the silicification of certain beds of limestone to a compact, conchoidally fracturing black jasperoid. The limestones are cut in all directions by veinlets of coarsely crystalline white calcite, and these were evidently formed at several distinct periods, since they cut and, in places, fault one another.

Quartzite forms a prominent bed near the middle of the Pogonip limestone, and in many places shale and quartzite beds are laminated with limestone near the top of the series. The quartzite is usually light in color and fine grained, although some beds are stained red by hematite and others are medium grained. Quartz veinlets with crystal-lined vugs cut the quartzite. Much of the quartzite is argillaceous and grades into slaty shales, which are thin bedded, fine grained, and, as a rule, olive green in color, although locally black or brown. Muscovite films are developed on the parting planes of both the shales and the more argillaceous quartzites.

This sedimentary series was evidently laid down in a sea of moderate depth which at times was shallow. The conditions were on the whole favorable to the deposition of limestone, although at intervals fine fragmental material was carried into that portion of the ancient sea now occupied by the Belted Range. This limestone closely resembles the Pogonip limestone of the Panamint Range, although from its great thickness it may extend well down into the Cambrian. Mr. E. O. Ulrich identified, from material collected by the writer, *Girvanella*-like forms of smaller size and denser structure than those collected from the Cambrian limestone of the Silver Peak Range. He considers these probably of Ordovician age.

Weber conglomerate.—The Weber conglomerate forms the crest and eastern slope of the Belted Range south of Whiterock Spring, as well as broad ridges west of Oak Spring and north of Whiterock Spring. The formation includes a sandstone from 800 to 1,000 feet thick and an overlying shale from 300 to 500 feet thick. While sandstone predominates in one member and shale in the other, some bands of shale are included in the sandstone, as well as sandstone in the shale member.

The reddish or brownish sandstone is either a quartzose rock or an arkose of medium grain. It is somewhat indurated, but when crushed fracture occurs around the grains rather than through them, as in quartzite. The indurated sandstone grades into conglomeratic bands which contain well-rounded pebbles up to 4 inches in diameter. These pebbles are smoky quartz of vein or pegmatitic origin, gray banded flint, pinkish or white quartzite, and black jasperoid seamed with quartz veinlets. The quartzite and jasperoid closely resemble the Cambrian rocks of other portions of the area, and if such is their age, it indicates that between the deposition of the Cambrian and the Carboniferous sediments a period of deformation intervened. The metamorphism of these supposed Cambrian pebbles is such that the beds from which they were derived must have been folded and eroded prior to the deposition of the Carboniferous strata. The arkose grades into thin-bedded shale of olive-green, brown, purple, or light-gray color.

The Weber conglomerate appears to lie conformably beneath the Pennsylvanian limestone, and hence occupies a position similar to that of the Weber conglomerate at Eureka, Nev., as described by Hague.[a] Near Whiterock Spring some peculiar fucoid-like markings occur on a light-gray shale. Mr. E. O. Ulrich states that these resemble *Palæodictyon* and are probably of Carboniferous age.

Pennsylvanian limestone.—Pennsylvanian limestone forms the prominent ridges north of Oak Spring and the low ridges south of the same point. It also covers considerable areas to the east and northeast of Tippipah Spring. Approximately 2,500 feet of this limestone is exposed in the Belted Range, although the top of the formation is not present. It is a fine- to medium-grained gray limestone of dense texture. Certain layers in fineness of grain and density resemble lithographic limestone. Much of it is magnesian, and other portions when struck with the hammer smell oily. Small black flint nodules occur in some beds, and white calcite veinlets are locally abundant. The limestone is for the most part rather heavily bedded, although interbedded shaly layers are fine bedded.

Intraformational conglomerates occur at several horizons, and near the center of the section as exposed is 55 feet of limestone conglomerate. The included pebbles, which reach a maximum diameter of 4 inches, are jasperoid and quartzite, probably of Cambrian age, and limestone, probably from the lower part of the Pennsylvanian section. It seems likely that this conglomerate merely indicates shallow-water conditions and not an important stratigraphic break, since Pennsylvanian fossils occur above and below it. Above this bed numerous thin layers of similar conglomerate and shale are interbedded with the limestone.

---

[a] Hague, Arnold, Mon. U. S. Geol. Survey, vol. 20, 1892, pp. 91–92.

The Pennsylvanian limestone appears to overlie the Weber conglomerate conformably. The formation is fossiliferous, and collections were made 4 miles south of Oak Spring and at several horizons northeast of Tippipah Spring. Dr. George H. Girty kindly made the following determinations of Pennsylvanian fossils, the first four lots being found beneath the limestone conglomerate and the fifth above it.

Fossils from Pennsylvanian limestone in Belted Range.

LOT 1.

Fenestella sp.   Productus sp.
Rhombopora sp.   Aviculipecten? sp.
Chonetes sp.   Phillipsia sp.
Productus cora?

LOT 2

Fusulina sp.   Marginifera? sp.
Rhombopora sp.   Seminula sp.
Archæocidras sp.   Aviculipecten sp.
Derbya? sp.

LOT 3.

Chonetes sp. aff. C. permianus.   Leda? sp.
Seminula sp.

LCT 4.

Zaphrentis sp.   Marginifera? sp.
Rhombopora sp.   Ambocœlia? sp.
Stenopora sp. cf. S. carbonaria.   Derbya? sp.

LOT 5.

Rhombopora sp.   Productus sp. cf. P. semireticulatus.
Stenopora? n. sp.   Productus nevadensis?

Siebert lake beds.—Siebert lake beds cover considerable areas north of Oak and Whiterock springs, and a number of small outliers occur on the Carboniferous rocks near by, indicating that the lake beds were at one time continuous with those of Pahute Mesa. Another outcrop lies south of Wheelbarrow Peak, and to the east there are smaller areas not mapped.

A section at Oak Spring is as follows:

Section at Oak Spring.

	Feet.
Later rhyolite	250
Well-bedded, incoherent sandstones and conglomerates; white with some dark gray bands. Many of the layers, which are from 2 to 4 feet thick, show cross-bedding. Crystals of glassy feldspar, quartz, and biotite are abundant in some beds and lacking in others. The pebbles in the conglomerate for the most part are small, but some of them reach 6 inches in diameter	200
Later rhyolite	40
White sediments as above	100
Later rhyolite	60
Salmon pink, white, and yellow sediments as above	280

Chalcedonic quartz, in considerable quantities, is deposited in the cavities of the sediments in this vicinity. The conglomerate near Wheelbarrow Peak incloses many bowlders of earlier rhyolite from 3 to 4 feet in diameter. Such bowlders protect the softer sandstone immediately beneath from erosion, and in consequence they form in places the caps of columns 20 feet high.

These beds appear to be contemporaneous with those of Pahute Mesa and Shoshone Mountain, and like them are considered the equivalent of the Siebert lake beds,[a] of Miocene age, at Tonopah.

IGNEOUS ROCKS.

Monzonite.—Included in the granite near Oak Spring are fragments of a fine-grained, dark-gray, granular igneous rock. On microscopic examination this proves to be a hornblende-biotite monzonite. Titanite and magnetite are accessory minerals.

Granite.—A stock of granite, approximately three-fourths of a mile in diameter, cuts the Pennsylvanian limestone 2½ miles south of Oak Spring and sends many apophyses into it. The granite, forming a dome, set with many exposures in blocklike masses, rises above the near-lying limestone. In the eastern portion of the area it has a sheeting dipping eastward, parallel to the limestone contact. The granite tends to weather into spheroidal masses, although some of the joint blocks have their corners but slightly blunted.

This rock is a light-gray porphyritic granite, characterized by unusually large feldspar and quartz phenocrysts, which lie in a medium- to coarse-grained matrix of white feldspar and biotite. Quartz is an unimportant constituent of the groundmass, the silica having largely separated out in the older crystals. Phenocrysts of pale-pink feldspar, of perfect outline, reaching a maximum length of 4 inches, are the most prominent, in places forming one-third of the rock surface. The feldspar phenocrysts inclose minute blades of biotite zonally arranged. The slightly smoky quartz phenocrysts are perhaps more numerous, though smaller than those of feldspar, their maximum diameter being one-half inch. Phenocrysts of mica 1 inch in diameter are rare. Both feldspar and quartz phenocrysts protrude on weathering and finally drop out. The granite is in places deeply weathered. The smaller feldspars are more kaolinized than the large crystals, while the biotites are bronze-brown in color and are surrounded by hematitic stain which has separated from them. Under the microscope this rock proves to be a biotite-hornblende granite, containing considerable plagioclase and hence leaning toward quartz monzonite. The texture is granular and hypidiomorphic, on account of the presence of plagioclase laths in the groundmass and of

---

[a] Spurr, J. E., Prof. Paper U. S. Geol. Survey No. 42, 1905, p. 51.

the large zonally built orthoclase phenocrysts. The accessory minerals include titanite, zircon, and apatite.

The granite is cut by dikes of aplite rather poor in biotite. Pegmatite dikes are common. The coarsely granular pegmatite is composed of feldspar and quartz anhedra which reach a maximum diameter of 12 inches. Limonite pseudomorphs after pyrite are inclosed in both quartz and feldspar grains. Graphic granite, with letters one-fourth inch across, is a less common variety of pegmatite. A third facies is an aggregate of coarse feldspar individuals with a few quartz anhedra. Embedded in the feldspar are numerous biotite blades from one-half to 1 inch long. Under the microscope this rock is seen to be composed of large plates of microperthitic orthoclase intergrown with quartz in the manner characteristic of graphic granite. Between these large areas is a granular mosaic of orthoclase and quartz. A few needle-like blades of biotite and grains of magnetite are also present.

Next to the limestone the granite becomes fine grained, although the phenocrysts retain their size and in consequence probably solidified prior to the intrusion of the magma into the limestone. Limonite cubes after original pyrite are particularly common near the contact. The limestone near the contact is metamorphosed to a fine-grained marble of white color, which in some localities contains lenses of pale-gray marble. In less pure layers fibrous tremolite, locally in globular masses, is developed. Other layers are colored greenish yellow by serpentine and chlorite. The more impure layers are changed to grossular garnet, forming a heavy dark-brown rock. This garnet in cavities occurs as beautiful citron-yellow crystals. A heavy rock of deep-green color also proves, on microscopic examination, to be composed of garnet, which even in thin section has a decided green tinge and which is associated with calcite, staurolite, tremolite, quartz, zoisite, titanite, and iron ore. Another thin section of altered limestone proves to be a finely granular marble, between the calcite grains of which are tremolite and colorless garnet grains and a few irregular areas of chlorite. Still another thin section is composed of large plates of calcite and quartz, in which chlorite, tremolite, and colorless garnet are micropoikilitically inclosed. The shale of the Weber conglomerate is metamorphosed to a slaty shale, in which nests of tiny muscovite tablets are developed. Other facies contain numerous pyrite crystals. The most interesting metamorphosed shale is a black argillite, in which are many rectangular white columns with black cores. Under the microscope these columns embedded in the carbonaceous argillite are seen to be chiastolite, a variety of andalusite. On its edges this mineral is somewhat altered to muscovite.

This coarse granite porphyry, characterized by two generations of mineral solidification and associated with aplitic and pegmatitic facies, clearly bridges the interval between granite and granite porphyry. The granite cuts Pennsylvanian limestone and is doubtless one of the post-Jurassic series. It rather strikingly resembles the granite outcropping 5 miles north of Ammonia Tanks.

Several smaller masses of similar granite occur near the main stock. The mass $1\frac{1}{2}$ miles east of south of Oak Spring lacks phenocrysts, and, since it is only 200 yards in diameter, is very similar in other respects to the contact facies of the main mass. It is medium grained or in a few places medium coarse grained. The granite is more altered than that of the main mass, and the mica is usually muscovite, apparently secondary to biotite. The granite grades into a coarse pegmatite, which in turn grades into pegmatitic quartz, each of which contains pyrite altered to limonite. The larger of the limonite cubes are one-half inch in diameter, and the pyrite was undoubtedly an original constituent of the pegmatite. The pegmatite has some miarolitic cavities into which quartz and feldspar crystals 1 inch long project. A third mass of granite lies between the two already described, and two small outcrops isolated in the alluvial deposits appear to have been separated from the largest mass by erosion.

Andesite.—About $2\frac{1}{2}$ miles west of Wheelbarrow Peak is an exposure of andesite partially covered by the earlier rhyolite flows. The rock has a dense reddish-black groundmass, in which are embedded many medium-sized striated feldspar phenocrysts, which are either glassy and colorless or cloudy and white. Hand specimens suggest the presence of hornblende. Under the microscope the groundmass appears as a glass containing many tiny laths of plagioclase. The phenocrysts are plagioclase (labradorite), brown hornblende, with a reaction rim of magnetite, which in some instances completely replaces the hornblende, and a little augite. The plagioclase phenocrysts, some of which are complex crystals, are somewhat rounded by magmatic corrosion and, as a rule, show beautiful zonal growth. Inclusions of a greenish-gray andesite, rather similar, altho with smaller feldspar phenocrysts, occur in rhyolite near Cliff Spring. In the gravels of the valley $1\frac{1}{2}$ miles west of Wheelbarrow Peak are bowlders of a greenish-gray andesite. Phenocrysts of altered greenish-white feldspar are abundant, and associated with these are biotite and either hornblende or pyroxene phenocrysts. The source of this andesite is unknown. It strikingly resembles the post-rhyolitic andesite of the Kawich Range and may perhaps be younger than the earlier rhyolite of the Belted Range. Certainly in this range andesite older than the rhyolites occurs, while andesitic eruptions may also have followed the rhyolitic effusion.

Earlier rhyolite.—The earlier rhyolite is the predominant rock of the northern part of Belted Range, and from its rather striking bands of red, purple, gray, and brown, the name of the range is derived. Of the many rhyolitic facies perhaps the most common has a rather dense reddish groundmass, in which are embedded medium-sized phenocrysts which equal or exceed it in bulk. The description of the phenocrysts of the earlier rhyolite of the Kawich Range applies almost equally well to those of this formation. Biotite, however, is the predominant phenocryst in a facies outcropping 1½ miles northwest of Cliff Spring. Hornblende phenocrysts are very rare. The rhyolite varies in color from white, gray, brown, or purple to .black. In texture the groundmass varies from glassy with well-developed flow lines, perlitic cracks, and spherulites through pumiceous to lithoidal. The rhyolites in the vicinity of Wheelbarrow Peak are more glassy and appear to be poorer in phenocrysts, particularly quartz. Chalcedony is abundart on joint planes, and throughout the area surveyed an unusually great development of this form of secondary quartz is associated with the more glassy facies. The only thin section examined proved to be a normal rhyolite with partially devitrified brown glassy groundmass.

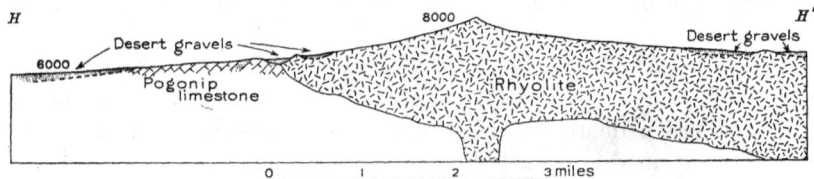

Fig. 12.—East-west section across Belted Range through Belted Peak.

Near Cliff Spring the red rhyolite of the cliffs has been bleached to a yellowish white by surface waters along joints and irregular cracks. The contact between the bleached and unbleached portions is unusually sharp. The joints are often offset regularly, and the bleaching follows such offsets. Microscopic examination shows that the bleaching is due to the removal of the hematite coloring matter of the somewhat devitrified glass. It is probable, in consequence, that considerable portions of the white chalky rhyolite of the range have been produced by similar bleaching.

Columnar jointing, due to contraction during cooling of the lava, is particularly well developed on a hill isolated in the wash, 1½ miles west of Cliff Spring, to the north of the road. The columns are from 6 to 12 inches in diameter and vary in position from vertical to horizontal, intermediate directions being common.

The rhyolite as seen on either flank of the range is clearly a flow, which in places is over 2,000 feet thick. Belted Peak has some of the characteristics of a vent (fig. 12), although the rhyolite may have been extruded from a long fissure near and parallel to the crest line, a suggestion which, however, requires for proof more careful field observation than was possible in the present reconnaissance. The relations between the small rhyolite mass 2 miles south of west of Belted Peak and the Paleozoic rocks are poorly exposed, but the rhyolite probably represents the lower portion of the rhyolite flow rather than a dike, since no dikes were observed in sedimentary rocks to the west.

The earlier rhyolite is similar lithologically to that of the Kawich and Reveille ranges, and like them lies beneath the Siebert lake beds. All three are practically of the same age, probably early Miocene.

Later rhyolite.—Later rhyolite caps Oak Spring Butte (see fig. 13) and forms a lower bench on the same butte and a similar bench on the ridge north of Whiterock Spring. It also covers large areas to the west, on Pahute Mesa. The predominant facies is a resinous or glassy rhyolite composed of light-gray and black flow bands and lenses. Perlitic parting is locally present. Medium-sized phenocrysts equal the groundmass in bulk and in the hand specimen consist of unstriated feldspar, often showing beautiful color plays, and colorless quartz in rounded grains or dihexagonal pyramid and prism crystals. The microscope shows the presence also of smaller phenocrysts of

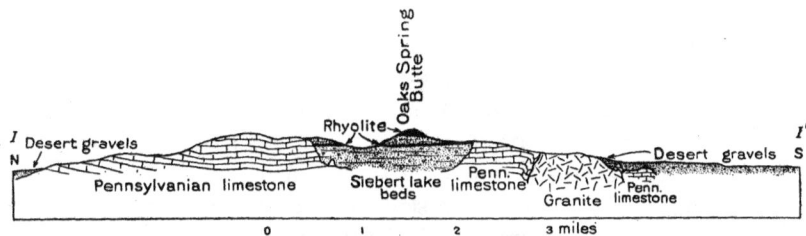

FIG. 13.—North-south section through Oak Spring Butte, Belted Range.

deep-green augite and biotite. Another facies is a reddish-gray aphanitic rock with a few phenocrysts of feldspar and black mica. Some quartz and plagioclase phenocrysts are also visible under the microscope in the devitrified-glass groundmass.

The later rhyolite is a flow, in part contemporaneous with the Siebert lake beds, but mostly younger. It is therefore of middle and possibly late Miocene age. It is contemporaneous with the rhyolite of Shoshone Mountain and Pahute Mesa, and probably with the later rhyolite of the Kawich Range and of Stonewall Mountain and with the later rhyolite and biotite latite of the Amargosa Range.

Basalt.—Basalt covers considerable areas near the middle of the Belted Range. The low hills at the north end of the range also appear from a distance to be basalt. The basalt, where examined, is lithologically similar to and probably contemporaneous with that of the Reveille Range, and, like it, overlies the eroded earlier rhyolite. It is probably of late Pliocene age. The surface of the Belted Range was, however, probably less rugged than that of the Reveille Range at the time of the basalt effusion.

STRUCTURE.

The Pogonip limestone, as exposed in the Belted Range, lies in a monocline which strikes north and dips 20° to 30° W. next Kawich Valley and 70° W. southwest of Belted Peak. Superimposed upon the main fold are a few minor isoclinal folds of similar strike, as well as gentle cross folds. This is probably the eastern limb of a broad syncline which underlies Kawich Valley and which has for its western limb the eastward-dipping monocline of Quartzite Mountain. A number of dip faults of east-west strike cut the Pogonip limestone, with a uniform offset of the beds on the north side of the faults to the east. The lateral displacement of the largest faults shown on the map is 100 feet.

The Carboniferous rocks are bent into rather open folds, which near Oak Spring have north-south axes and to the north of Tippipah Spring, as a rule, course northeast and southwest. Small isoclinal folds passing into overthrust faults occur, particularly in the shale. The intrusion of the granite considerably disturbed the strata on its north side, while the beds on the other sides were but little affected. The folding was evidently in part prior to the granite intrusion, in part due to it, and in part later, since the granite in places has a sheeting parallel to the bedding planes of the limestone. Since the folding of the strata some normal faults have been formed.

Prior to the extrusion of the earlier rhyolite in the northern part of the range the Pogonip limestone formed a mountain range possibly somewhat lower than the present Belted Range, with its crest 3 miles west of the present crest line. In the southern part of the range prior to the deposition of the Siebert lake beds the Carboniferous rocks had at least a gently accentuated surface, since numerous hillocks of the older rocks protrude through the younger sediments north of Tippipah Spring. The Siebert lake beds at Oak Spring have been uplifted without important flexure. After this uplift erosion developed a mature mountain surface, which prior to the extrusion of the basalt was tilted to the east. The basalt flows on the east side of the range dip to the east at the rate of 500 feet to the mile, possibly indicating that the eastward tilting continued after the outflow of the basic lava.

ECONOMIC GEOLOGY.

OAK SPRING.

At Oak Spring a number of prospects are being developed. In granite 1½ miles nearly due south of Oak Spring quartz veins of pegmatitic origin, from 1 to 3 feet wide, striking N. 30° E. and dipping 15° NW., have been staked. The quartz is white and slightly sugary and contains vugs with small quartz crystals. Some of it is intensely brecciated, the cracks being stained by hematite and limonite. Sulphides are sparingly present and consist of pyrite, chalcopyrite, galena, and zinc blende, named in the order of their abundance. From these hematite, limonite, malachite, azurite, and cerussite are derived as secondary minerals. A coating of an unctuous mineral in silvery tablets frosts some of the cavities. The values in this locality are said to be gold with less silver. The deposits are genetically similar to the mineralized pegmatite veins of Lime Point, in Slate Ridge, and the sulphides were probably introduced into the pegmatitic quartz after its solidification.

The pegmatitic quartz veins abundant in a small granite mass three-fourths of a mile east of south of Oak Spring contain pyrite crystals, as do the less acidic pegmatites and the surrounding and genetically related granite. The quartz is said to carry good gold and silver values, presumably in the pyrite.

About 300 yards southwest of Oak Spring and down the same ridge is a 25-foot shaft in Pennsylvanian limestone, here locally horizontal. A vein 2 feet wide, which strikes N. 35° E. and dips 70° NW., cuts the limestone. The vein is formed of malachite, chrysocolla, and a jaspery quartz, which is deeply stained by blotches of manganese dioxide and limestone. Striking vugs in malachite lined with later azurite which in turn is covered with clear quartz crystals resemble copper-sulphate crystals. Massive yellowish-gray cerussite (lead carbonate) is also present. A yellowish-green, finely scaly coating on fractures was determined by Mr. Waldemar T. Schaller to be either emmonsite or durdenite, hydrated ferric tellurites. The presence of the compound of tellurium is of considerable interest, since it indicates the presence of a telluride among the original minerals of the post-Jurassic ore deposits. Emmonsite is one of the last secondary minerals to form and coats both jaspery quartz and chrysocolla. Postmineral faulting has occurred. The secondary minerals, which alone are seen, partly replace the limestone and partly fill preexisting cavities.

The so-called turquoise mine at Oak Spring is a small cut in the metamorphic Pennsylvanian limestone $1\frac{1}{2}$ miles south of the spring and three-fourths of a mile from the granite contact. Two veins strike north and south and dip 65° W., apparently parallel to the bedding of the limestone, which is here partially marmorized and silicified. The wider vein varies in width from 2 inches to 1 foot and can be traced several hundred feet. Four feet away is a parallel vein, the two being connected by a few chrysocolla stringers. The veins are composed of a mottled mosaic of chrysocolla and a dark compact jaspery quartz, stained in some places by limonite and in others by manganese dioxide. The chrysocolla is usually verdigris green, although picked pieces are a beautiful robin's-egg blue. The substance is usually opaque, but some is slightly translucent. The chrysocolla is commonly massive, locally with a botryoidal structure, and in some massive phases an occasional cleavage face is seen. It is cut by veinlets of manganese dioxide and white calcite or quartz. Chrysocolla veinlets of slightly different color cut one another, showing that the formation of this mineral extended over a considerable period. Associated with it is a crystalline, bottle-green, semi-transparent mineral whose cleavage faces reach a length of one-half inch. A radial structure is observed in places. This mineral is embedded in the chrysocolla or cuts it in veins and is of practically contemporaneous age. It is probably brochantite, a hydrous sulphate of copper. Both the chrysocolla and the brochantite were determined by Mr. Waldemar T. Schaller. Postmineral faults cut the veins and parallel them. Pieces of blue chrysocolla closely resemble turquoise, and several hundred pounds of the material have been sold for this gem. The mineral takes an excellent polish. The largest piece of pure chrysocolla seen was 6 by 3 by 2 inches.

In these copper deposits the minerals exposed are all secondary. Malachite, chrysocolla, brochantite, cerussite, a jaspery quartz, and limonite seem practically contemporaneous. There is considerable evidence that the formation of the copper minerals and the jaspery quartz extended over a considerable period, during which some fracturing occurred and in consequence the relations between these minerals are complex. They partly replace the limestone and partly fill fissures. Azurite, emmonsite, quartz, and calcite are of later origin.

Narrow seams and veins of quartz are rather widely distributed in the Pennsylvanian rocks, particularly near granite intrusions. In the slaty shales and quartzites throughout the southern portion of the range such veins occur. A quartz vein 2 inches thick which carries some pyrite cuts the shale one-half mile south of Whiterock Spring. Specimens of quartzite, which evidently came from the quartzite to the north, were found at Tippipah Spring. These were rather strongly stained by malachite and hematite.

Oak Spring furnishes sufficient good water for domestic purposes, and there are several other springs in the general vicinity. Oak Spring Butte is timbered to some extent. Calientes, on the San Pedro, Los Angeles and Salt Lake Railroad, is the natural shipping point.

OTHER AREAS.

Prospects are located on either side of the road from Kawich to Cliff Spring, near the eastern border of the Pogonip limestone. The supposed ore is in part a fine-grained quartzite and in part vein quartz, in which are small disseminated iron-pyrite cubes and thin stringers of pyrite. The veins where examined are thin.

Reefs of silicified rhyolite similar to that of the Kawich Range occur south of Belted Peak. The alteration of the rhyolite is strikingly like that at Silverbow, Eden, and other mining camps situated in rhyolite, and the ground is worthy of prospecting. Over a considerable area east and southeast of Belted Peak, near the boundary of the area mapped, the rhyolite appears in a distant view from its color and topography to be much altered.

PAHUTE MESA.

TOPOGRAPHY AND GEOGRAPHY.

The name Pahute Mesa is applied to the lava-capped table-land stretching from Stonewall Mountain on the northwest to the Belted Range on the southeast, a distance of 48 miles. The mesa is 12 miles wide except near its center, where Tolicha Peak cuts it into two portions. Notwithstanding this break these divisions are so similar, both topographically and geologically, that it is convenient to describe them under a single name.

Pahute Mesa consists of a series of benches one above another, the surfaces of which are determined by resistant flows of lava (Pl. II, A). These level benches are trenched deeply by canyons, while hills of older rocks and cones, the vents of the lavas, rise from their surfaces. Southeast of Stonewall Mountain the mesa has an average elevation of 5,500 feet. It gains in altitude to the southeast and near the Belted Range the highest mesa bench is 7,500 feet high. The southward-facing scarp is 1,500 feet high.

The western portion of the mesa is without vegetation other than sage and related shrubbery and the yucca. The higher portions of the eastern edge are, on the other hand, covered with a fairly heavy growth of piñon and juniper, and grass grows luxuriantly in the higher valleys, making this portion of Pahute Mesa the best winter range in the area mapped. The mesa has no large springs, Pillar Spring containing less than 50 gallons of water. A well sunk in a canyon 3 miles north of east of Gold Center yields about 300 gallons of water a day. It is 90 feet deep and passes through Recent gravels into tuffaceous sediments beneath the basalt. Tanks are common on the surface of the mesa, but in summer, with one probable exception, these are dry. A little bad water can be procured, probably throughout the year, by digging in the gravels at Ammonia Tanks. On account of the lack of water Pahute Mesa is more dangerous in summer than Death Valley.

GENERAL GEOLOGY.

On account of its inaccessibility and the comparative simplicity of its geology, but little time was devoted to the mapping of this region. While many inaccuracies exist in the geologic boundaries as here laid down, they broadly represent the main features. The latest rhyolite and later tuffs beneath the basalt in the northwestern portion of the mesa are not differentiated from the basalt, while the separation of the Siebert lake beds and the middle rhyolite near the Belted Range is only approximate.

The list of the formations of Pahute Mesa, from the oldest to the youngest, is as follows: Mica schist, pre-granite monzonite, post-Jurassic granite, earliest rhyolite, biotite andesite, Siebert lake beds, middle rhyolite, latest rhyolite, later tuffs, and basalt.

METAMORPHIC ROCKS.

Mica schists.—At Trappmans Camp the granite contains fragments of mica schist, and an outcrop several hundred yards in diameter, 1 mile east of Trappmans Camp, may extend eastward a considerable distance beneath the Tertiary lavas. These rocks are crenulated and foliated schists. One facies is a dark-brown schist apparently composed of tiny plates of biotite and muscovite well arranged in the plane of foliation. The microscope shows that the two micas are distributed in rather distinct bands parallel to one another. Andalusite, sillimanite, and finely divided magnetite are present as accessories. Another facies is a silvery schist composed of muscovite, with here and there an ellipsoidal aggregate of deep-green chlorite. Under the microscope this proves to be a schist composed of the essential constituents, muscovite, chlorite, and quartz, and the accessory constituents, sillimanite, zircon, andalusite, and rutile. Nothing is known concerning the age of these intensely metamorphosed rocks beyond the fact that they are much older than a granite of supposed post-Jurassic age. They are perhaps of Cambrian age, since they somewhat resemble the more metamorphosed Cambrian schists near granite at Gold Mountain.

SEDIMENTARY ROCKS.

Paleozoic quartzite.—Prospectors report that quartzite occurs on the north slope of Mount Helen, but the locality was not visited by the writer.

Siebert lake beds.—The Siebert lake beds west of the Belted Range are similar to those described in the vicinity of Oak Spring. (See p. 122.) The beds are either horizontal or tilted up to angles of 30°. Probably to be correlated with these are the white conglomeratic sandstone 10 miles north of Tolicha Peak, and green tuffaceous sediments 7 miles southwest of Trappmans Camp. From the tuffaceous beds, which contain quartz, feldspar, and biotite crystals, Mr. B. D. Stewart collected pieces of silicified wood which Dr. F. H. Knowlton states belonged to a deciduous tree not older than the Tertiary.

Later tuffs.—Beneath the basalt of Pahute Mesa east of Stonewall Mountain are tuffs and rhyolite flows which appear to be similar to

the later rhyolite and later tuffs of the Goldfield hills. (See pp. 72, 74.)
Similar beds outcrop at many points where erosion has removed the
basalt, and it is probable that these beds underlie the greater portion
of the basalt. No attempt has been made to separate them from
basalt on the map. A section east of Stonewall Mountain is as
follows:

Section east of Stonewall Mountain.

	Feet.
Basalt.	
Rhyolite	5
Gray tuff with small pebbles	15
Rhyolite	7
Yellow tuff, very rich in pebbles of pumice, which reach a maximum diameter of 1 inch	20
Gray, rather incoherent, conglomeratic tuffaceous sediments. The pebbles, which reach a maximum diameter of 6 inches and become smaller toward the top, are formed of rhyolitic pumice, a quartz-poor rhyolite like the later rhyolite of Goldfield, and basalt. Tiny crystals of feldspar, similar to the phenocrysts of the rhyolite of Goldfield, are present	55
Yellow tuffs, like above	10

From this section and that at Goldfield it may be inferred that a
lake covered considerable territory in the northern part of the area
under discussion, probably in late Pliocene time. The deposits of this
lake are separated from those of the Pahute Lake (Siebert lake beds)
by an erosional unconformity, and are probably older than the playa
deposits of the older alluvium in Stonewall Flat. Between the Pliocene lake beds and the older playa deposits there is little relation,
although the deposits laid down in playas probably represent the last
stages of the drying up of the lake in which the lake beds were
formed. Spurr[a] found deposits of a late Pliocene lake widely distributed in Nevada, and it is possible that this lake and the one here
described are broadly contemporaneous.

Recent desert gravels.—Recent desert gravels mask the bed rock
over a considerable area east of Fortymile Canyon and extend some distance up the main canyons.

IGNEOUS ROCKS.

The igneous rocks of Pahute Mesa include pre-Tertiary granular
rocks and Tertiary dikes and lava flows.

Pre-granite monzonite.—In the granite 5 miles north of west of
Whiterock Spring are inclusions of a gray fine-grained granular rock.
Both striated and unstriated feldspars are present with quartz and
biotite, and the rock is similar to the monzonite included in the post-
Jurassic granitoid rocks of Gold Mountain and the Belted and Pana-
mint ranges.

Post-Jurassic granite.—Granite forms the hills at Trappmans Camp
and also outcrops 5 miles north of west of Whiterock Spring. The

---

[a] Spurr, J. E., Bull U. S. Geol. Survey No. 208, 1903, pp. 124–125, 209–210, etc.

rock at Trappmans Camp is a light-gray fine- to medium-grained biotite-muscovite granite. It forms low, rounded hills and shallow valleys, and outcrops are inconspicuous. The granite is cut by several systems of intersecting joints and by faults and zones of brecciation. In consequence the residual masses are squared blocks, more or less rounded. Under the microscope the granite shows as an allotriomorphic, granular rock, composed of orthoclase, quartz, plagioclase, biotite, muscovite, and zircon. The muscovite is in part original and in part an alteration product of feldspar. The rock has been considerably mashed, and quartz and muscovite have been deposited in the fractures of feldspar and quartz.

The granite grades into and is cut by irregular masses and dikes of pegmatite, composed of feldspar and quartz anhedra up to 1 inch in diameter. Quartz stringers and veins of pegmatitic origin, some of which are several hundred feet long and 40 feet wide, are common. The quartz on portions of the borders of the larger masses contains fragments of granite and sends well-defined veins into the granite. At other places single individuals of quartz on the contact seem common to both pegmatitic quartz and granite, and arms of feldspar extend from the granite into the quartz. Evidently the quartz solidified contemporaneously with some portions of the granite and subsequently to other portions. A few thin dikes of fine-grained aplite, which stand in relief on weathering, cut the granite.

A granite mass 1 mile long occurs in the midst of the Siebert lake beds 5 miles north of west of Whiterock Spring. It is low and covered by a yellowish granite soil from which protrude low, rounded granite domes. Near its center is a rugged hill, set with sharp pinnacles, whose forms are controlled by the well-developed joints. The granite is coarse grained and is composed of pink feldspar, white semitransparent quartz, and biotite. Feldspar and quartz reach a maximum length of three-fourths of an inch. Under the microscope the granite shows as a hypidiomorphic granular rock formed predominantly of large orthoclase plates poikilitically inclosing quartz anhedra and plagioclase laths. A little biotite is also present, and this, since it occurs along fractures in orthoclase, has been considerably recrystallized. The accessory minerals are titanite, apatite, and magnetite, the last probably titaniferous, since it is surrounded by secondary titanite.

The granite is cut by narrow dikes of pink aplite, practically lacking biotite. These dikes weather in relief. In the granite are ellipsoidal masses of quartz-feldspar pegmatite from 4 inches to 5 feet in diameter, the contact between the granite and pegmatite being in some instances sharp, in others gradational. Feldspar and quartz individuals with a maximum diameter of 6 inches, together with a few biotite plates, compose the pegmatite. This rock grades into pegmatitic quartz. The ellipsoidal form and the absence of channels from one mass to another in the place of observation suggest that the pegmatite formed in place from the local residual fluids of the granitic magma.

The granite of Trappmans Camp contains inclusions of the schist already described and of a black Paleozoic quartzite. It is cut by dikes of the earlier rhyolite, and the later tuffs overlap it. It resembles the granite of Lone Mountain. The granite near Whiterock Spring is much older than the Siebert lake beds and closely resembles that near Oak Spring in the Belted Range. Each granite is probably of post-Jurassic and pre-Tertiary age.

Earliest rhyolite.—The earliest rhyolite protrudes from the basalt in a number of hills in the northwestern portion of the mesa. It evidently occurs for the most part as a flow, although dikes of rhyolite cut granite at Trappmans Camp. These dikes reach an observed maximum width of 40 feet. They are formed of a lithoidal, brownish-gray rhyolite with conchoidal fracture. The feldspar, quartz, and biotite phenocrysts rarely exceed a length of one-tenth of an inch and are exceeded in bulk by the groundmass. The flow at Wilsons Camp is composed of a white phenocryst-rich rhyolite, in which biotite is present in some places and absent in others. The rock of the inlier, $9\frac{1}{2}$ miles northwest of Tolicha Peak, is a semipumiceous, brownish-gray rhyolitic glass, with a few small phenocrysts. Pebbles of black obsidian occur on the mesa near by. Other hills in the vicinity were not visited, but on the basis of their color and the report of prospectors some are mapped as rhyolite. Beyond the fact that the rhyolite of this vicinity is younger than the granite and older than the basalt, its age is unknown. It is believed, however, that the different masses are portions of the rhyolite of the Cactus Range buried by later lava flows, and are in consequence of early Miocene age.

Biotite andesite.—Biotite andesite forms the inlier of Gold Crater and several smaller inliers near by, and also occurs as dikes cutting the rhyolite at Wilsons Camp and the granite at Trappmans Camp.

The biotite andesite at Gold Crater forms a group of low hills which protrude above the basalt, later rhyolite, and later tuffs. It is evident that the andesite flow was eroded into hills prior to the formation of the younger rocks, and it is probable that the younger rocks never completely covered the summits of these hills, although liquid lava dammed up against their sides. The biotite andesite here is for the most part intensely altered, but appears to have been originally a dense gray rock with phenocrysts equaling the groundmass in bulk. The phenocrysts are medium sized and consist of predominant feldspar in laths, biotite, and locally hornblende. Flow banding and flow breccias are present. Under the microscope the microfelsitic groundmass is rather ill defined, but seems to be composed predomi-

nantly of plagioclase and orthoclase. The phenocrysts include sericitized feldspars, which were at least mainly plagioclase, biotites
altered to muscovite and magnetite, and some hornblende-like forms of
serpentine and limonite.

The biotite andesite at Wilsons Camp is a highly altered, finely
granular rock of pale-greenish color, containing many phenocrysts
of white altered feldspar and some of bronze biotite. Probably to be
correlated with this rock is a much altered greenish-gray, finely granular rock which occurs in dikes at Trappmans Camp. Laths of altered
feldspar, columns of hornblende, and tiny tablets of biotite are de-
terminable in the hand specimen. The microscope shows that the
much altered groundmass was probably originally a glass with a few
plagioclase laths (hyalopilitic). The only phenocrysts recognizable
are large calcitized plagioclase and chloritized biotite and a few
small, fresh orthoclase tablets. Aggregates of secondary minerals
suggest the presence of hornblende or pyroxene. Apatite and magnetite, probably titaniferous, are accessory minerals.

The andesite at Wilsons Camp bears the same relation to the rhyolite
as the biotite andesite of the Cactus Range, and the two are doubtless contemporaneous and of middle Miocene age. The andesite of
Gold Crater is much older than the latest rhyolite. Lithologically it
is not very different from the biotite andesite and is provisionally correlated with it. It is, however, possible that it is the flow equivalent
of the monzonite porphyry of the Kawich Range.

Middle rhyolite.—The middle rhyolite, evidently a flow, is the predominant formation of Pahute Mesa east of Thirsty Canyon. The
most widely distributed type is a gray or pink lithoidal rhyolite, in
which the groundmass is exceeded in bulk by the medium-sized phenocrysts. The latter include glassy orthoclase, colorless or slightly
smoky quartz, and, usually, biotite. This variety in many places
contains small fragments of a white pumiceous rhyolite and in consequence is a flow breccia. Other facies include dense rocks of red,
brownish, or gray color, with few phenocrysts. Flow banding, spherulites, and perlitic parting accompany the denser groundmass in
much of the rock. Black glasses with feldspar phenocrysts occur
near Ammonia Tanks, and gray semipumiceous glasses with abundant
feldspar and quartz phenocrysts are found on the east side of Silent
Canyon, 4 miles south of its mouth. This rhyolite for the most part
lies upon the Siebert lake beds and it is to be correlated with the later
rhyolite of the Belted Range. It is presumably of late Miocene or
early Pliocene age.

Latest rhyolite.—The latest rhyolite occurs below the basalt in the
vicinity of Stonewall Mountain and probably outcrops in many other
places on the border of the mesa that are shown on the map as basalt.
It is interbedded with and overlies the later tuffs. A section already

described (see p. 133) shows that there are two thin flows of this rhyolite. The lower flow is a dense grayish-purple rock containing a vast number of small phenocrysts. Under the microscope the groundmass is seen to be a glass with highly developed flow lines. The phenocrysts all appear to be orthoclase. The upper rhyolite is similar to the lower, although it contains slightly larger orthoclase phenocrysts, and in addition a few tiny biotite and quartz phenocrysts are visible. The younger rhyolite of Pahute Mesa and that of Goldfield hills are similar lithologic units occupying like stratigraphic positions, and without doubt are contemporaneous and probably of late Pliocene age.

Basalt.—Flows of basalt cover Pahute Mesa northwest of Tolicha Peak, similar flows extend eastward from Quartz Mountain to Thirsty Canyon, and a third area of considerable size lies west of Timber Mountain. The last-named mass appears from a distance to have flowed from two cones, one 5 miles northeast and the other 8 miles southeast of Timber Mountain. Black Mountain, from a distance, also resembles a volcanic cone, and the north peak of Mount Helen may have been an ancient vent.

The basalt of Pahute Mesa includes vesicular and nonvesicular facies, the former red, the latter black. In some of the basalt white striated feldspars, reddish altered olivines, and black pyroxenes are prominent, while in other portions these phenocrysts are inconspicuous. Stratigraphically the basalts of Pahute Mesa and of the Goldfield hills have identical positions and they are considered of late Pliocene age.

STRUCTURE.

Pahute Mesa northwest of Tolicha Peak is capped by lava flows, which dip gently in all directions from a center 5 miles southwest of Mount Helen. The only structural feature of interest in this portion of the mesa is the slight rise of the lava sheets as they approach the older hill groups, well seen at Gold Crater and on the southeastern face of Stonewall Mountain. A similar gentle dome is present in the mesa between Quartz Mountain and Thirsty Canyon, but eastward from the canyon the mesa gradually rises, owing probably to comparatively recent uplifts of the Kawich and Belted ranges. In this portion of the mesa normal faults with north-south strike, few of which have a displacement of over 50 feet, are common. Although of late Tertiary age, these faults do not exhibit fault scarps, since erosion has planed down the upthrown side. In some instances, however, the downthrown side, being more resistant, is now a scarp.

ECONOMIC GEOLOGY.

Three mining camps are situated in Pahute Mesa—Trappmans Camp, Wilsons Camp, and Gold Crater.

TRAPPMANS CAMP.

Trappmans Camp lies 34 miles south of east of Goldfield. The veins were discovered by Hermann Trappman and John Gabbard in June, 1904, and at the time of the writer's visit a year later five men were opening up the veins, the chief development being a shaft 50 feet deep.

The prospects are in granite, and three distinct periods of vein formation were noted—first, quartz lenses probably of pegmatitic origin; second, quartz veins of distinctly later formation, which in one place are said to cut a rhyolite dike; and, third, quartz veins of a third generation which cut the second. (See fig. 14.) The latest veins are in places well crustified.

The pegmatitic quartz forms bodies varying from minute stringers in the granite to lenses one-fourth mile long and 40 feet wide. The quartz is hard and whitish, and in some places it is intensely

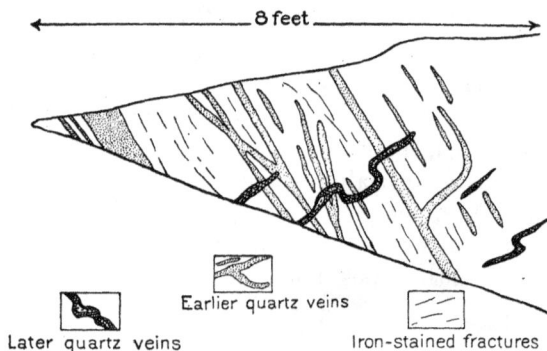

FIG. 14.—Second and third systems of quartz veins in granite at Trappmans Camp, as exposed on face of incline.

brecciated and stained by limonite. It is said to carry silver and gold values.

The quartz veins of the second class have sharp contacts with the granite. These veins are exceedingly common in the vicinity of Trappmans Camp and vary widely in strike and dip. In limited areas they tend to form series of veins along parallel joint planes in the granite. In width the veins vary from an inch or less to a foot or more, and some of the parallel series are a number of feet thick. Locally these veins are faulted, as indicated by the presence of breccias, while the surrounding granite shows considerable differential movement. The quartz is slightly bluish, but is usually heavily stained red or brown by hematite and limonite. Vugs elongated parallel to the direction of the veins are sometimes seen. On encountering the pegmatitic lenses, in the one instance noted, these veins are deflected downward.

The veins of the third class are of later origin than those of the second, which they cut. As a rule they are more narrow and less continuous and dip more steeply than the veins of the second class. In places they curve sharply. The quartz in them is similar to that in the second group.

In the vicinity of the vein the granite is cut by partings stained by limonite and manganese dioxide, which are parallel to or join the quartz veins at low angles, and the surrounding kaolinized granite is said to carry values. The granite near the veins is in places thickly peppered with cubes of pyrite altered to limonite.

The ore of the two younger sets of veins so far encountered is practically all oxidized, although a little original galena remains. The predominant original sulphide was pyrite, and limonite cubes after pyrite are common in the quartz. Assays show the values to be in the proportion of one of gold to four of silver. Some silver chloride was noted, while secondary native silver is reported. In one prospect calcite is associated as a gangue with quartz.

While the quartz of the first set of veins is of pegmatitic origin, its mineralization is probably later and genetically connected with the filling of the fissures of the second set, probably in Tertiary time. After the veins of the second system were fractured another period of mineralization followed, perhaps in late Tertiary time. Later the veins were crushed and surface waters have more or less completely oxidized the sulphides.

Wood and water are hauled from Antelope Springs, about 9 miles away. Trappmans Camp is 40 miles by road from Goldfield.

WILSONS CAMP.

Wilsons Camp is 2 miles north of Trappmans Camp and was discovered in May, 1904. Five miners were employed in July, 1905, and at that time several shallow shafts and short tunnels were open. The country rocks, white altered rhyolite and biotite andesite, are cut by rather steeply dipping quartz veins, the majority of which strike northeast, although some strike east. The quartz veins, many of which are crustified, are characterized by quartz-lined vugs. Since its formation the quartz has in some instances been crushed. Limonite and less commonly malachite stain the quartz. The reported assay values run from $110 to $180 per ton and average one of gold to five or six of silver. These quartz veins are to be correlated with the veins of Silverbow. The economic conditions at Wilsons Camp are similar to those at Trappmans.

GOLD CRATER.

The mining camp of Gold Crater is situated 10 miles east of the summit of Stonewall Mountain. The first locations were made in

May, 1904. In the fall of 1904 several hundred people rushed to the camp, but few remained long. At the time of the writer's visit (July 7, 1905) a number of lessees were at work. The country rock, silicified and kaolinized biotite andesite, has been fractured and in many places faulted and brecciated. The intensely silicified andesite is a white or gray rock with conchoidal fracture. In some instances it is porous through the removal of the phenocrysts and in others the casts have been filled by milky quartz. The iron-stained outcrops are very rugged, since silicification and consequently the resistance of the rock to erosion are very irregular throughout the mass. Under the microscope this silicified facies is seen to be a medium to very fine mosaic of quartz and some chalcedony, with here and there a blotch of limonite. No phenocrysts remain, although a few sagenitic webs of rutile suggest the former presence of biotite. By kaolinization the rock is reduced to a chalky mass, in which biotite phenocrysts, altered to a silvery mica, alone are visible. It is intensely stained by limonite and hematite, especially along fractures, and from such places rich gold pannings are obtained. A little chrysocolla was observed along some joints, while a thin coating of hyalite has been deposited since the oxidation of the sulphides. The ore is said to run from $40 to $240 in gold per ton.

Waters carrying silica and metallic salts in solution appear to have ascended along faults, brecciated zones, and joints in the country rock and to have deposited silica, pyrite, and some copper sulphide. Later surface waters oxidized the sulphides and set the gold free. The original deposition was without much doubt an impregnation of the country rock, as is the case with the secondary minerals. There is a notable resemblance between these deposits and those of Kawich and certain of those of Goldfield.

Gold Crater derives its water supply from tanks on the basalt mesa and from two wells, 3 and 9 miles distant. Fuel for mining purposes is obtainable from Stonewall Mountain. Goldfield is 27 miles distant.

OTHER AREAS.

Some of the pegmatitic quartz in the granite 5 miles north of west of Whiterock Spring is stained by limonite, but no other indication of mineralization was noted. It is probable that some of the masses of the older Tertiary volcanic rocks protruding through the later lava flow are worthy of prospecting, but such were not observed in the course of the present work. Prospecting in Pahute Mesa can be most advantageously pursued in the winter or early spring, when water can be obtained from snow or the tanks.

TOLICHA PEAK AND QUARTZ MOUNTAIN.

TOPOGRAPHY AND GEOGRAPHY.

Tolicha Peak, a striking cone rising from Pahute Mesa, is a landmark for miles. Quartz Mountain, a somewhat lower east-west ridge, extends eastward from Tolicha Peak. The hills are bare, except for a sparse growth of desert shrubbery. Monte Cristo Springs, on the western slope, furnish sufficient water for 15 to 20 head of stock.

GENERAL GEOLOGY.

The formations of these mountains from the base up are Cambrian (?) schist, post-Jurassic granite, Tertiary rhyolite, and Tertiary basalt.

SEDIMENTARY ROCKS.

Cambrian (?) schist.—The hill 2½ miles southeast of Tolicha Peak appears greenish gray in color at a distance, and is probably formed of a bluish-gray schistose rock which occurs abundantly in the desert gravels to the south of this hill. The rock is very fine grained, except for some small crystals of biotite and hornblende. White, ellipsoidal areas 1 inch long spot the blue-gray rock. Under the microscope the groundmass appears as an exceedingly fine aggregate of quartz and feldspar, in which are embedded small areas of quartz, orthoclase, ragged hornblende, and biotite. The rock is probably a metamorphosed Paleozoic shale, possibly of Cambrian age.

IGNEOUS ROCKS.

A specimen of rhyolite from Tolicha Peak collected by Mr. S. G. Benedict contains an inclusion of gray granite of medium grain. Post-Jurassic granite probably underlies this peak.

Rhyolite forms Tolicha Peak and the east side of Quartz Mountain. The predominant type is a dense flinty rock of reddish color with a few small phenocrysts of glassy feldspar, quartz, and biotite. Flow lines, perlitic parting, and spherulites are well developed. Gray glassy facies, transparent in thin flakes, are interbedded with this type, as are white or light-gray incoherent facies in which the phenocrysts are abundant and equal the groundmass in bulk. Flow breccias are rather common. This rhyolite is probably about contemporaneous with the earlier rhyolite of the Kawich Range and is presumably of early Miocene age.

Prior to the extrusion of the basalt which covers Pahute Mesa (see p. 137) Tolicha Peak was eroded into a rugged mountain. The valley between Tolicha Peak and Obsidian Butte is upon the contact between the rhyolite and basalt.

The rhyolite of Tolicha Peak is complexly cut by normal faults with displacements of 10 to 50 feet. These faults in some places are less than 100 yards apart. The rhyolite is also intensely jointed.

The rhyolite around Monte Cristo Springs is kaolinized or silicified. Prospect holes are located on quartz veins in altered rhyolite, the ore being quartz and silicified rhyolite stained by limonite. Quartz Mountain Camp, on the northern slope of Quartz Mountain, was not visited, but the conditions, according to the reports of prospectors, are similar to those at Monte Cristo Springs. The altered masses, which resemble those of Silverbow and are faulted like those of Bullfrog, are outlined on the economic map (fig. 4, p. 43).

MOUNTAINS SOUTHEAST OF PAHUTE MESA.

Southeast of Pahute Mesa are a number of small groups of mountains with crest lines of varying trend, although a number have an eastward extension. These include Shoshone, Skull, Yucca, Timber, and Bare mountains and the Specter Range. Geologically they are characterized by the important development of the early Miocene rhyolite.

Shoshone and Skull mountains lie to the north of the Specter Range and the Skeleton Hills and to the east of Fortymile Canyon. Shoshone Mountain is joined to the Belted Range by a low gap, but from its east-west trend it more properly belongs with the other ridges in the vicinity. Shoshone (7,540 feet) and Skull (6,100 feet) mountains are each capped by horizontal lava flows, and in consequence from a distance appear as mesas. The other hills and ridges vary in form from the jagged ridges of Carboniferous rocks to the smooth depressed domes of the Siebert lake beds. The various ridges and hills are in many instances separated from one another by opposed valleys which are filled with desert gravels.

A fairly heavy growth of juniper and piñon is found on the north slope of Shoshone Mountain above Tippipah Spring, but other portions of this mountain and the other ridges are bare except for sparse groves of the tree yucca and here and there a Spanish bayonet (*Yucca aloifolia?*). Scarcely a spear of grass grows throughout these hills. Of the four springs in these hills, Cane Spring flows about 1,500 gallons a day; Tippipah contains from 50 to 100 gallons of standing water; Topopah (or Blackrock) flows from 15 to 25 gallons a day, and the fourth, which lies 4 miles south of Cane Spring, is small and is dry in summer.

GENERAL GEOLOGY.

The formations of these hills, from the oldest to the youngest, are the following: Carboniferous (Pennsylvanian) sedimentary rocks, quartz-monzonite porphyry, earlier rhyolite, Siebert lake beds, later rhyolite, and basalt.

SEDIMENTARY ROCKS.

Pennsylvanian limestone.—Pennsylvanian limestone forms the ridge to the northeast of Tippipah Spring and the northern face of the ridge northwest of Yucca Pass. A third area of limestone, 6 miles south of Tippipah Spring, outcrops between the desert gravels below and the later rhyolite above. Limestone pebbles are abundant in the Siebert lake beds, and small inclusions occur in the quartz-monzonite porphyry and the earlier rhyolite. This limestone is similar to that of the Belted Range. (See p. 121.)

Siebert lake beds.—Siebert lake beds form the northern and northeastern faces of Shoshone Mountain, rim the southern edge of Skull Mountain, and cover areas 4 miles southeast of Tippipah Spring and 4 miles south of Cane Spring. Somewhat over 1,000 feet of lake beds are exposed on the north side of Shoshone Mountain, a section being as follows:

Section on north side of Shoshone Mountain.

	Feet.
White (locally yellow through iron stains), fine-grained, incoherent, impure sandstone. Bedding planes near the base 100 feet apart. Crystals of biotite, glassy feldspar, and quartz occur in upper 200 feet. Some bands contain pebbles of glassy rhyolite less than 3 inches in diameter.	650
Same, but without the small crystals. Some pink layers. Bedding planes well developed from 40 feet to 1 foot or less apart. Pebbles of glassy rhyolite up to 3 inches in diameter occur in certain beds	100
Same, with pebbles of Carboniferous limestones and quartzites one-fourth inch in diameter	200
Same, with abundant crystals of biotite, glassy feldspar, and quartz. Pebbles up to 4 inches in diameter, largely of glassy or semipumiceous rhyolite	100

In other areas these sediments are more brilliantly colored, pinks and reds not being unusual. The bowlders of the conglomeratic beds in some instances reach a diameter of 5 feet. In addition to the pebbles already mentioned, others are present of quartz-monzonite porphyry, the earlier rhyolite, and a vesicular basalt. The basalt was not seen in place, but is perhaps contemporaneous with the earlier rhyolite and corresponds to the earlier basalt of the Amargosa Range. The Siebert lake beds, which characteristically form depressed domes, break down readily into a sand. Huge globular cavities, due in part to exfoliation and in part to wind erosion, and pillars capped and protected by huge bowlders add fantastic elements to many hills. Joints affect only the minor topographic features. In certain instances silicification has followed joints and chalcedony has been deposited in their fractures. Such joints, being resistant, weather in relief.

The Siebert lake beds appear to have been deposited in a lake of some depth, to which much fragmental material was carried. The size of the material varied considerably from time to time, the variation in coarseness being an expression in part of the depth of the lake and in part of the topography of the lake shores. Much of the material was derived from disintegrating rhyolite, but at times there were apparently also explosive rhyolitic eruptions, which threw into the lake many small crystals of biotite, feldspar, and quartz. The sediments on Shoshone Mountain overlie the quartz-monzonite porphyry and earlier rhyolite unconformably and are overlain apparently conformably by later rhyolite, while on Skull Mountain the basalt overlies them unconformably. They are without much doubt to be correlated with the Siebert lake beds at Tonopah [a] of Miocene age.

IGNEOUS ROCKS.

Quartz-monzonite porphyry.—Quartz-monzonite porphyry is the predominant formation in the hills to the east, north, and northeast of Cane Spring. The texture of the rock indicates that it was probably intruded in older rocks, which are, however, not at present exposed.

The quartz-monzonite porphyry has a gray or greenish-gray, finely crystallized or lithoidal groundmass, which is somewhat subordinate in bulk to the phenocrysts. The latter include feldspar, usually altered but in rare instances fresh and then clearly striated; biotite, fresh to altered, and serpentinized hornblende. In many cases hornblende is very subordinate to biotite. The feldspars reach a maximum length of one-fourth inch; the other phenocrysts are slightly smaller. The microscope shows that the rock is a quartz-monzonite porphyry almost approaching granodiorite porphyry. The groundmass consists usually of a microgranitic, but over small areas micropegmatitic mosaic of orthoclase, quartz, and a little plagioclase. The plagioclase phenocrysts, complex crystals, are usually twinned according to the Carlsbad law and many of them are zonally built. One determination proved this mineral to be basic andesine. A few phenocrysts of quartz, deeply embayed by magmatic corrosion, and of orthoclase are also present, in addition to the phenocrysts microscopically determined. Magnetite, zircon, and apatite are present as accessory minerals. The secondary minerals include kaolin and epi-

---

[a] Spurr, J. E., Prof. Paper U. S. Geol. Survey No. 42, 1905, p. 54.

dote after the feldspars, chlorite and epidote after biotite, and pseudo-morphs of serpentine after hornblende or possibly augite.

The rock resembles lithologically and in its stratigraphic position the monzonite porphyry of the Kawich Range, and the two rocks are doubtless approximately contemporaneous. Each is strikingly similar to the quartz-monzonite porphyry of Stonewall Mountain. The porphyry of Skull Mountain and that of the Kawich Range are older than a rhyolite, which is supposed to be of early Miocene age. The quartz-monzonite porphyry also closely resembles rocks which on the western border of the area surveyed appear to be a variant of the post-Jurassic granite magma. While it is possible that they are pre-Tertiary in age they are here tentatively assigned to the late Eocene.

Earlier rhyolite.—The earlier rhyolite and its kaolinized and iron-stained alteration products form the Calico Hills, remarkable for their brilliant pink, red, and white coloring. A number of smaller areas lie in the group of hills in which the Horn Silver mine is situated, and some of the rhyolite to the west of Yucca Pass may possibly belong to the earlier series.

The earlier rhyolite includes a number of facies, most of which are characterized by abundant phenocrysts that equal or exceed the groundmass in bulk. The groundmass is white or gray in color and is usually dense. The phenocrysts are of medium size. Tabular crystals of glassy unstriated feldspar and rounded grains of slightly smoky quartz are much more abundant than the somewhat smaller hexagonal plates of black mica. Other facies are black glasses with perlitic parting, but without phenocrysts, and gray glasses with a few quartz phenocrysts. The rock is a flow, since certain beds show distinct flow banding and others contain many fragments of rhyolites and are evidently flow breccias.

The earlier rhyolite appears to lie upon the uneven surface of the quartz-monzonite porphyry. The later rhyolite apparently overlies the earlier, while pebbles of the earlier rhyolite are contained in the Siebert lake beds. The same relations exist in the Kawich Range, and the earlier rhyolites of the two ranges are doubtless approximately contemporaneous and of early Miocene age.

Later rhyolite.—The later rhyolite flow overlies the Siebert lake beds on Shoshone Mountain, while several small masses to the south of Cane Spring cap the same formation. In the latter locality the former extension of the rhyolite was much greater and the south-ward-dipping flow of the Shoshone Mountain may at one time have been connected with it. The rhyolites west of Yucca Pass are, from their lithologic character and relation to the Siebert lake beds, probably to be correlated with the later rhyolite.

Southwest of Tippipah Spring 275 feet of the later rhyolite over-lies the Siebert lake beds, while near the center of Shoshone Mountain about 1,500 feet of rhyolite is exposed. At the first locality the lower 250 feet is a dense brownish-pink rock with sparse and small phenocrysts of glassy feldspar, biotite, and quartz. Some flow bands are purple or gray in color, others are vesicular, and still others carry numerous small inclusions of rhyolite, and are in consequence flow breccias. Vertical columnar joints, originating during the cooling of the magma, are not unusual. The broadest columns are 4 feet in diameter. Above this is 15 feet of a black glass, in which the phenocrysts exceed the groundmass in bulk. This in turn is overlain by 10 feet of rhyolite like that first described, except that the phenocrysts are larger and more numerous. In it are thin bands of the black glassy variety. At the south end of Shoshone Mountain similar facies occur, and in some of these the flow banding is strongly developed, the flow lines, as a rule, being straight, although in places wavy.

The later rhyolite south of Cane Spring is a yellowish-brown rock with lithoidal or chalky groundmass, which is equal in bulk to the phenocrysts. These include medium-sized glassy unstriated feldspar and slightly smoky quartz, together with abundant but tiny biotite phenocrysts. The rhyolites west of Yucca Pass are waxy-lustered glasses and well-banded, phenocryst-rich flow breccias of purplish color. In the latter gray spherulites occur.

The later rhyolite ordinarily forms buttes and mesas, since the various flow bands vary considerably in resistance to erosion. Like the later rhyolite of the Kawich Range, the later rhyolite of these hills overlies the Siebert lake beds. The two, which are rather similar lithologicaly, are probably about contemporaneous and are of late Miocene or early Pliocene age.

Basalt.—A basalt flow caps Skull Mountain, a number of small areas occur to the east of the Calico Hills and to the west of the Horn Silver mine, while three small hills lie 4 miles east of the iron tank in Fortymile Canyon. Basalt also occurs between Fortymile Canyon and Shoshone Mountain. It is a dense black or dark-gray rock with sparse phenocrysts of glassy striated feldspar and slightly altered glassy olivine. Both the vesicular and nonvesicular facies previously described in other ranges occur. The nonvesicular basalt weathers into spheroidal masses, while the vesicular facies usually break down into slabs which are elongated parallel to the flow banding. A single thin section of a dense, slightly vesicular facies proved on microscopic examination to be composed of dark glass containing many laths of a calcic plagioclase and grains and partial crystals of olivine more or less altered to reddish-brown serpentine.

On Skull Mountain the basalt lies upon the eroded surface of the Siebert lake beds, while in the low isolated hills 4 miles east of the iron tank it appears to overlie the later rhyolite. It is without doubt the youngest formation of these hills, and is probably of late Pliocene or early Pleistocene age.

STRUCTURE.

The Pennsylvanian limestone is folded into rather open folds, the dips of which rarely exceed 45°. The predominant axes trend from northeast to southwest, although minor folds with east-west axes also occur. The ridge to the northeast of Tippipah Spring is synclinal in structure. The Pennsylvanian rocks appear to have had a rather uneven surface prior to the deposition of the Siebert lake beds. Not only do many of the lake beds contain large bowlders of limestone, but west of Tippipah Spring a number of hillocks of the older limestone protrude through the Tertiary sediments.

In a broad way the Siebert lake beds and the later rhyolite are slightly flexed and the beds on Shoshone Mountain dip gently to the south. Normal faults of 50 feet displacement are not unusual in the tuffs. The latest deformation is that of the basalt. On Skull Mountain it dips gently to the south, while on the ridge to the east it dips with a similar angle to the north, the two blocks being separated by a north-south fault.

ECONOMIC GEOLOGY.

The quartz-monzonite porphyry and the earlier rhyolite have been silicified and kaolinized in restricted areas, the silicification being comparable to the alterations in the vicinity of Kawich and the kaolinization to those of Silverbow. These areas, which are worthy of prospecting are delineated in fig. 4 (p. 43). The quartz-monzonite porphyry near the Horn Silver mine has not only been altered to a spongy mass of brownish-gray silicified porphyry like that of Kawich, but it is cut by quartz stringers with associated gypsum. Other veins and irregular cavities are partially filled with chalcedony, which in turn is incrusted by quartz crystals. The rhyolite of the Calico Hills has been in places intensely silicified. The silicified rhyolite 2 miles north of the Horn Silver mine is cut by thin quartz seams. A similar yellow stain (a basic ferric-alkali sulphate?) to that noted near the prospects at Silverbow is present on joint surfaces.

SPECTER RANGE AND SKELETON HILLS.

TOPOGRAPHY AND GEOGRAPHY.

The hills in the southeast corner of the area surveyed may be considered as the westward extension of the irregular Spring Mountain Range. They consist of a number of mountains and hills now isolated from one another by narrow canyons and broad valleys filled with desert gravels. The ridges and hills are rugged and sharp and are crowned by many peaks. The highest point (5,800 feet above sea level) is in the southeast corner of the area. The presence of many inlying hills near by indicates that those portions of the old mountain mass partially buried by desert gravels possessed an equally dissected topography. A majority of the hills are elongated parallel to the strike of the Paleozoic rocks. West of the road between Cane Spring and Ash Meadows ridges with northeasterly trend predominate, and east of the road lie irregular massive groups of hills. The hills stop abruptly along a north-south line east of Fortymile Canyon, and prior to the inwash of detrital material this canyon must have been a strikingly prominent topographic feature. The hills are bare of timber and destitute of springs.

GENERAL GEOLOGY.

The formations of these hills, from the oldest to the youngest, are Prospect Mountain quartzite, Prospect Mountain limestone,[a] and basalt.

SEDIMENTARY ROCKS.

Prospect Mountain quartzite.—Two areas of quartzite, one large and the other small, cover 20 square miles near Point of Rocks. The formation, 2,000 to 3,000 feet thick, underlies the Prospect Mountain limestone conformably, with transitional beds between the two. The rock, either a quartzite or an indurated sandstone, varies in color from gray to purplish red and in grain from fine quartzite to conglomerates, with pebbles one-fourth inch in diameter. The pebbles include white quartzite, red and brown jasperoid, and white, smoky, or opalescent quartz of vein or pegmatitic origin. The rocks are rather constantly impure and grade through arkoses into minor beds of red and green slaty shale. The coarse-grained arkose resembles granite on first sight. The quartzite is usually considerably jointed.

This rock appears to have been deposited in a comparatively shallow sea, to which much fragmental material was carried. From its position, conformably beneath Cambrian limestone, this quartzite is undoubtedly to be correlated with the Prospect Mountain quartzite,[b] of lower Cambrian age.

Prospect Mountain limestone.—The Prospect Mountain limestone is the predominant rock of the Specter Range and the Skeleton Hills. It is typically light to dark gray, compact, crystalline, and fine grained. Some beds have a conchoidal fracture and are almost as compact and fine grained as lithographic limestone. Pinkish gray varieties of medium grain also occur, but are rather unusual. Cross-bedding is observed in places. The beds are typically massive, although certain horizons are thin bedded. On weathering the various beds

---

[a] See note. p. 28.

[b] Hague, Arnold, Mon. U. S. Geol. Survey, vol. 20, 1892, p. 35.

assume red, yellow, and gray colors of varying tints, rendering the mountain structure legible at great distances. Near the middle of the section, as exposed, are a few layers of black chert, while a bed of quartzose sandstone, 20 to 30 feet thick, occupies a similar position in the more northerly hills east of Fortymile Canyon. This rock, locally a quartzite, is white or pinkish white and fine grained.

The formation appears to be between 5,000 and 6,000 feet thick. It was deposited in a sea of medium or shallow depth, in which depression of the sea bottom kept pace with deposition and into which little detrital material was introduced. A few poorly preserved fossils were collected north of the Skeleton Hills. Mr. E. O. Ulrich reports that these consist of trilobite fragments and a valve of a brachiopod, probably referable to the genus *Billingsella*. He believes that they are of Cambrian age. The limestone is without much doubt the Prospect Mountain (Cambrian) limestone of Eureka, Nev.,[a] although the upper portions may belong more properly with the Pogonip limestone.

IGNEOUS ROCKS.

In no other equal area surveyed during this reconnaissance do igneous rocks cover so little territory. The writer saw no indications

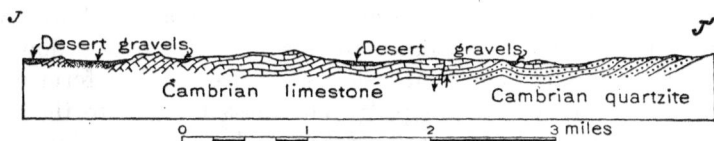

FIG. 15.—Northwest-southeast section through Specter Range.

of granite or other intrusions, an observation verified by statements of several prospectors who have examined the hills in considerable detail. A single small butte of vesicular basalt similar to that of the mesa south of Cane Spring protrudes from the desert gravels northeast of Point of Rocks. This butte is probably an erosional remnant of a flow which extended from this mesa southward.

STRUCTURE.

The Cambrian rocks are folded in a complex manner. Two sets of folds are present in almost equal development, the axes of one striking northeast, those of the other striking east and west. Zones of brecciation and normal faults occur at a number of places, but few of the faults show a displacement of more than 30 feet. The fault 2 miles north of west of Point of Rocks is normal and has a displacement of 100 feet. It trends N. 30° E. and the western side is downthrown. A sag in the hills coincides with the fault, the downthrown side being slightly higher than the upthrown. Fig. 15 is a section passing through the hills from northwest to southeast.

---

[a] Hague, Arnold, Mon. U. S. Geol. Survey, vol. 20, 1892, pp. 36–38.

ECONOMIC GEOLOGY.

Although the Cambrian rocks are cut by thin veins of white crystalline or banded calcite and by thin quartz veins, apparently barren, they appear to offer slight attractions to the prospector. The Johnnie mine, where strong quartz veins occur, lies 2 miles south of the area surveyed. From information furnished by prospectors it is probably in the Prospect Mountain quartzite.

YUCCA MOUNTAIN.

TOPOGRAPHY AND GEOGRAPHY.

Yucca Mountain is so called from the groves of the tree yucca which cover its lower slopes. The group lies between Fortymile Canyon on the east, Beatty Wash on the north, Oasis Valley and Crater Flat on the west, and the Amargosa Desert on the south. The mountains of the main east-west ridge are gently sloping domes with rugged summits, the highest of which is 6,700 feet above sea level. The ridge drops rather suddenly to Fortymile Canyon on the east, but on the west it descends gently and merges with Bare Mountain. From the main east-west ridge mesa tongues extend southward to the Amargosa Desert, rhyolite being the predominant rock. Yucca Mountain is colored dull tones of gray, brown, red, and yellow. Near the summits the varying resistances to erosion of the rhyolite flow beds are expressed in cliffs and gentle slopes. Spires tend to form where vertical joints are well developed. The mesa tongues to the south become lower toward the south-southeast, the surface being determined by resistant lava flows. The eastern side of these mesa tongues passes beneath the Recent alluvial deposits, while the western face is, as a rule, a scarp. The magenta-red basaltic cone 4 miles east of Rose's Well is a prominent landmark. Other volcanic cones are located in Crater Flat and for convenience will be described with Yucca Mountain.

The wide and open character of Beatty Wash between Timber and Yucca mountains is noteworthy. Alluvial deposits cover the depression for a width of 2 miles, although many hills of solid rock protrude from them. The narrowness of the canyon farther downstream is due partly to the more resistant character of the rhyolite to the west. while the width of the upper valley is probably due to the rapid waste of the adjoining high mountain masses on either side.

No water is known in Yucca Mountain. A few junipers and piñons grow on the highest summits of the main ridge.

GENERAL GEOLOGY.

This mountain is composed entirely of igneous rocks, which include the following, beginning with the oldest: Post-Jurassic pegmatite, rhyolite and latite, and basalt.

Post-Jurassic pegmatite.—A single fragment of quartz-feldspar pegmatite was noted in the basalt of the cone 8 miles north of Rose's Well. The basalt in its passage to the surface evidently passed through a mass of pegmatite.

Rhyolite and latite.—The mountain mass, with the exception of a number of rather small basalt areas, is composed wholly of rhyolite and latite. The series includes siliceous volcanic rocks of varying aspect. The brown groundmass in many instances is dense, aphanitic, and opaque, and from the presence of flow banding, perlitic parting, spherulites, and lithophysæ, is evidently of a glassy nature. Gray and black semitransparent glasses also occur. The more altered facies have a gray or white dull lithoidal groundmass and the microscope proves that this also was once a glass that is now devitrified. The phenocrysts are medium sized and usually subordinate to the groundmass; in certain instances they are practically lacking. Slightly smoky rounded quartz individuals and glassy unstriated feldspar crystals are present in equal development; smaller black or bronze biotite crystals are less abundant. In some facies the phenocrysts are all orthoclase, but microscopic examination shows enough quartz in the groundmass to determine their rhyolite nature.

Latites[a] are interbedded with the rhyolites. They are slightly darker in color and contain both striated and unstriated feldspar phenocrysts in approximately equal development. The single specimen examined microscopically proves to be a clear glass, the brownish color of the hand specimen being due to abundant curved trichites radiating from a common center. Perlitic parting is beautifully developed. Phenocrysts of orthoclase and plagioclase of medium composition are each more abundant than biotite. Wedge-shaped titanite crystals, magnetite, and a little hornblende are also present.

Interbedded with the rhyolite and latite flows are porous semi-tuffaceous facies. Flow breccias also occur, the inclusions being of a slightly different rhyolite than that of the main mass. The central portion of the main east-west ridge is probably the site of a vent from which siliceous volcanic rocks flowed. The flows, which extended particularly to the south and west, reach a minimum thickness of 1,000 feet. The series is a continuation of similar rocks of the Bare Mountain group. Presumably it is to be correlated with the rhyolites of the Bullfrog Hills and the Kawich, Belted, and Reveille ranges, and is probably of early Miocene age.

Basalt.—In Crater Flat are two large and one small volcanic cone and a number of low mesa ridges of basalt. A fourth cone is situated

---

[a] The use of the term latite is here tentative, since no chemical analyses of the rocks have been made.

4 miles east of Rose's Well. The basalt of the cones is vesicular and dull brick-red or black in color, while in the flow pediment around the cones are dense black facies with a few greenish glassy grains of olivine.

The cones are gently depressed and are usually superimposed upon circular basalt flows. The flows present ropy surfaces, cross fractures produced during the flow of the almost solid mass, and caverns formed by the onflow of the liquid interior after the surface had hardened into a crust, each a characteristic phenomenon of surface lavas which have been but little eroded. The cones themselves are formed in part of flows, but largely of vesicular lapilli, scoriæ, and volcanic bombs. The intimate mixture of these red and black fragments imparts to the cones their magenta-red color. The cone 4 miles east of Rose's Well has on its summit a crater, 300 feet in diameter, which is depressed from 15 to 75 feet below its rim. Sulphur coats the volcanic breccia in the crater. The crater of the cone 7 miles north of Rose's Well has been largely destroyed by erosion, the throat displaying a rubble of vesicular basalt fragments. Beds of fragmental material dip inward toward a common center at the crater of the cone 1½ miles farther north.

Since the eruption of the basalt the cones have been somewhat eroded, calcium carbonate has been deposited in the vesicles, shrubbery has taken root on the lava flows, and a sand dune has been superimposed upon the cone 4 miles east of Rose's Well. The basalt is probably of very late Pliocene or early Pleistocene age.

STRUCTURE.

The rhyolite-latite series has been subjected to mountain-building forces acting in an east-west direction. Joint sheeting and normal faults striking north and south are developed in the mass, while minor joints of east-west strike are also present. The series had been carved into forms approximating those of the present day when basalt eruptions occurred.

ECONOMIC GEOLOGY.

Two small masses of rhyolite whose location is shown on the map (fig. 4, p. 43) are kaolinized and silicified in a similar manner to the rhyolite in the vicinity of ore deposits. Silicification has occurred along joints and in irregular forms, and these resistant masses form the topographic details peculiar to silicified areas of rhyolite. Tunnels have been driven in the larger of the two areas, the ore extracted being a silicified rhyolite slightly stained by limonite. The ore, in part free milling and in part refractory, is said to carry gold with minor silver values. Quartz does not seem abundant in either area.

TIMBER MOUNTAIN.

TOPOGRAPHY AND GEOGRAPHY.

Timber Mountain, named on account of its heavily forested summits, is bounded on the east and northeast by Fortymile Canyon, on the northwest and west by Oasis Valley, and on the south by Beatty Wash. A low divide connects it with Pahute Mesa to the north. The massive mountain has two peaks, each 7,500 feet high, which are separated by a col. The western slope to Oasis Valley is somewhat more gentle than that on the east to Fortymile Canyon. The heaviest forest growth in the area surveyed covers this mountain above an elevation of 6,300 feet. Some of the trees are of sufficient size to be of value as timber.

GEOLOGY.

Timber Mountain appears to be wholly composed of Tertiary rhyolite and basalt, the rhyolite being the older.

Rhyolite.—The mountain peaks and the north and west slopes are formed of rhyolite, which imparts to the mountain a red tone. The most abundant variety is a dense brown or red rock in which the phenocrysts, medium-sized glassy orthoclase and slightly smoky quartz with fewer bronze biotite, equal or exceed the groundmass in bulk. The rhyolite is but an extension of that of Yucca Mountain, and the various facies there noted probably also exist on Timber Mountain. The rhyolite is presumably of early Miocene age.

Basalt.—The rhyolite of Timber Mountain had been eroded into practically its present form when basalt was extruded, filling depressions in the rhyolite and forming level reaches here and there on the mountain. Basalt flanks Timber Mountain on the east, forms the prominent black dome 8 miles east of south of the more northerly of the two peaks, and occurs in several areas on the western slope of the range. The basalt is a dense black or dark-gray rock in which striated feldspar and olivine phenocrysts are prominent. Much of the basalt is vesicular, and the rounded cavities, locally filled by zeolites, vary greatly in size across the flow banding. The basalt flow reaches a thickness of over 1,000 feet. The basalt of Timber Mountain is much younger than the rhyolite, and from the amount of erosion which it has suffered it is believed to be one of the late Pliocene or early Pleistocene basalts.

BARE MOUNTAIN.

TOPOGRAPHY AND GEOGRAPHY.

Bare Mountain, so named on account of its almost total lack of verdure, lies south of Oasis Valley and Beatty Wash, between the Amargosa Desert and Crater Flat. The mountain mass is triangular in form, the apex being at the south end. It is 15½ miles long from north to south and 8 miles wide at the north, its broadest portion. The Bare Mountain hills are comparable to the most rugged portions of the Amargosa and Panamint ranges, the valleys being deep V-shaped troughs and many of the ridges between being knife-edges. Throughout the range the south sides of the gulches are somewhat less steep than the north sides, owing to the prevalent dip to the northeast. The highest point of Bare Mountain, 6,235 feet above sea level, is but 2 miles from the detrital pediment 3,000 feet below. The strongly folded Paleozoic rocks band the mountain with dark and light gray, yellow, white, and red. No water is known in these bare hills.

GENERAL GEOLOGY.

The formations exposed in Bare Mountain, from the oldest to the youngest, are as follows: Pogonip limestone, Eureka quartzite, pegmatite, monzonite porphyry, rhyolite, and basalt.

PALEOZOIC SEDIMENTARY ROCKS.

Bare Mountain is composed largely of Paleozoic sedimentary rocks so strongly folded and faulted that a satisfactory section could not be made in the time at the writer's disposal. The series presented from the top down is probably as follows:

Probable section in Bare Mountain.

	Feet.
Schist	200
Quartzite with some schist and limestone	1,000
Limestone containing near its top a considerable number of beds of schist	3,000–4,000
Quartzite and schist	100–200

At the top of the series the schist, limestone, and quartzite repeatedly replace one another within a thickness of 100 feet. These rocks, particularly at the north end, have been metamorphosed, the original limestone, shale, and sandstone being altered, respectively, to marble, schist, and quartzite.

The limestone in the central part of the range is dark gray to blue-black in color, fine grained, compact, and rather crystalline. Anastomosing veinlets of calcite are characteristic. Much of the limestone is thinly laminated. At the north end of the mountain the limestone is changed to a white or light-colored marble, for the most part rather coarse grained. Weathered surfaces are yellow or brownish in color. Muscovite films are developed on parting planes. The upper and lower quartzites consist of a white or light-gray metamorphosed quartzose sandstone of fine or medium grain and conchoidal fracture. Cross-bedding is common and ripple marks are present in places. In transition to schist the quartzite becomes darker in color, finer in grain, and less pure, some muscovite being present. The original shale of the series, now a silvery or greenish-gray mica schist, is well exposed near Gold Center. The schist is a fine-grained, foliated, and locally crenulated rock composed of micaceous minerals with some quartz and feldspar. Small magnetite and pyrite crystals are common in certain bands and are evidently contemporaneous with the other metamorphic constituents. The planes of schistosity are usually parallel to the bedding planes, which are in places ripple marked, although in certain instances the foliated structure crosses the original bedding at right angles. Weathered outcrops have a typical rusty appearance. A single thin section under the microscope showed a thoroughly recrystallized rock lacking all traces of original structure. The rather well oriented micaceous minerals are muscovite, biotite, and limonite-stained chlorite, and grains of quartz and orthoclase lie between their blades. Tiny magnetite crystals and grains are abundant.

The lower limestone member of the series was deposited in a sea of medium depth, into which for a long period little or no fragmental material was carried. The upper portion of the series, on the other hand, was deposited in a shallow sea which at times received fragmental material of varying grain. The quartzites and schists mark periods in which the deposition of fragmental material exceeded organic and chemical sedimentation, while the interbedded limestones indicate a reversal of the dominant process. A few poorly preserved fossils (one of which is identified by Mr. E. O. Ulrich as a gasteropod suggesting *Eccyliopterus* or some other euomphaloid genus) were obtained from Bare Mountain, and although by no means diagnostic of the age of the series, they are probably younger than the Cambrian and older than the Carboniferous. The basal quartzite may be an intercalated bed in the Pogonip limestone or possibly may represent the Hamburg shale of the Eureka section. The thick limestone on lithologic grounds is correlated with the Pogonip (Ordovician) limestone, while the overlying quartzite is probably the Eureka (Ordovician) quartzite of Hague.[a] This upper quartzite and schist rather closely resemble the Eureka quartzite of the Kawich Range. The area covered by the two formations is roughly indicated on the map (Pl. I), but future work will prove the distribution much more complex.

IGNEOUS ROCKS.

Post-Jurassic pegmatites.—Pegmatitic dikes from 2 inches to 4 feet thick are common in the northern part of the range. The most widely distributed type is composed of white or very light pink feldspar, with some colorless quartz and muscovite. None of the individuals exceed a diameter of one-half inch. A less common variety

---

[a] Hague, Arnold, Mon. U. S. Geol. Survey, vol. 20, 1892, pp. 54-57.

is a glassy quartz in which are many biotite flakes one-eighth inch in diameter. A few black tourmaline rods up to one-fourth inch long occur in a quartz-rich muscovite pegmatite. While some of the tourmaline occurs in nests of clustered rods, some also forms along minute cracks and appears to be of pneumatolitic origin. Pegmatitic quartz veins with small quantities of feldspar and muscovite are very common, and it is probable that many of the older quartz veins were deposited by waters from the granite magma. The pegmatite dikes cut Paleozoic rocks and are included in the Tertiary volcanic rocks, and are probably post-Jurassic in age. They closely resemble the pegmatites west of Bullfrog.

Diorite porphyry.—Bowlders of the biotite-monzonite porphyry or diorite porphyry occur on an alluvial fan at the northwest edge of Bare Mountain. Dikes of this rock, possibly of pre-Tertiary age, probably cut the Paleozoic rocks.

Tertiary rhyolite and basalt.—Flows of rhyolite and basalt form the north end of the Bare Mountain hills. These are similar lithologically to the rhyolite and later basalt of the Bullfrog Hills. (See pp. 180–181.)

STRUCTURE.

The Paleozoic rocks of Bare Mountain form a faulted monocline, the beds of which strike from north of west to northwest and dip from 15° to 80° NE. Superimposed upon the monocline are minor cross folds. The schist and in a less degree the limestone are closely crenulated and small isoclinals and overturned folds are common. The folding was so intense that schist bands between limestone and quartzite are severed on the limbs of the folds and greatly thickened at the crests and troughs. At least two profound normal faults of over 1,000 feet displacement cut the Paleozoic rocks, and other faults of almost equal magnitude exist. Small faults, both normal and reverse, are common. Joints traverse all the formations, but are particularly developed in the quartzite, which in consequence breaks down into squared blocks.

Prior to the extrusion of the Tertiary rhyolite the range stood out boldly and was probably as high or almost as high as at present. The differential movement which has occurred at the contact of the Paleozoic and Tertiary rocks is unimportant.

ECONOMIC GEOLOGY.

Quartz veins are common in the north end of the range, and mineral locations were made on some of them early in 1905. Older veins, which were folded with the Paleozoic rocks, are cut and faulted by a younger set. All the quartz veins are characterized by rather abrupt variations in width.

The Decillion claim is located in the northwestern portion of the range near the mouth of a deep canyon. The principal development work is a 50-foot shaft. Quartz veins and stringers reaching a maximum thickness of 1 foot cut limestone and schist. Some of the veins strike east and west and dip 45° S. Chalcopyrite in grains up to one-half inch in diameter is disseminated through the quartz. Malachite and a little azurite stain the quartz around the sulphide and have been deposited in cavities in the quartz. The quartz vein since its deposition has been faulted and crushed, the fragments being in places cemented by granular gypsum, while gypsum crystals line some cavities. The ore of this mine is said to run high in silver, with low gold values.

The two 50-foot shafts of the Kismet Mining Company, better known as the Lonsway property, are situated on the crest of a high ridge 1½ miles southwest of Beatty: The country rock here, a quartzite, strikes N. 35° W. and dips 15° NE. The quartz vein, which is traceable for several hundred feet, is from 1½ to 2 feet wide and strikes N. 85° W. and dips 55° S. The quartz has a sparry texture and is white except at the surface, where it is heavily stained by limonite and hematite. This vein is said to have been locally charged with gold, probably derived from pyrite which occurs with limonite in quartz at the bottom of the shaft. The quartzite near the vein is sheeted. Silver ores with purple-fluorite gangue occur at the Lige Harris prospect. In other portions of the range the quartz veins contain galena cubes.

The quartz and associated sulphides were deposited in open east-west fissures, possibly by granite magmatic waters. Later the veins were faulted and crushed, and surface waters have more or less completely altered the sulphides to oxides and carbonates, forming free-milling ores.

The mines on Bare Mountain are controlled by practically the same economic conditions as those of the Bullfrog district. Good water can be obtained in the springs of Amargosa River one-half mile away. The most accessible wood is that of the Amargosa Mountains, 25 miles distant. The mines are 1 to 3 miles from the proposed railroad to the Bullfrog district.

SARCOBATUS FLAT.

Sarcobatus Flat lies west of Pahute Mesa and east of the Amargosa Range and Slate Ridge. It is one of the largest valleys in the area surveyed, being 32 miles long and having an average width of 10 miles. It has very gently graded slopes. Near its center, at an elevation of 3,950 feet, is a large playa, and two small playas lie in the arm of the valley south of Tolicha Peak. The surface of the larger playa is slightly roughened by erosion, but the total difference of altitude over the entire flat is probably less than 20 feet. A few lime-carbonate concretions of rude cylindrical form are embedded in the playa clay. Similar concretions occur in the playa clay of the Alkali Spring flat.

Adobe bricks have been made from the playa clay at several places, and it is probable that the clay of all the playas is well adapted to their manufacture. Borax is said to have been leached from the clay of this flat 3 miles south of east of Montana Station. The industry was never extensive, and until the rich bedded deposits of borax minerals are exhausted these playa deposits will be commercially unimportant. A number of wells, sunk in or near the large playa, have uniformly obtained water. In the center of the playa the surface and the water table are practically at the same level (see fig. 2, p. 21). The coyotes scratch holes from 2 to 3 feet deep in the clay and get water. The water is adequate for domestic and stock purposes and for the most part pure, although some is brackish and that in one well is unfit for use. Water was obtained at the following depths in the various wells: Forks of Goldfield-Bullfrog and Goldfield-Beatty roads, 180–190 feet; Farmer Station, 8–10 feet; Thorp's mill, 30 feet; Montana Station, 18 feet; Summerville, 1 mile south of Montana Station, 8–9 feet; Tonopah Well, 10–12 feet, and Seattle Well, 40–50 feet. Mr. E. L. Rickard reports that the water at Montana station occurs in a layer of quicksand.

AMARGOSA DESERT.

The Amargosa Desert is one of the most extensive depressions in the Great Basin. The main valley heads against the Amargosa Range and Bullfrog Hills, south of Sarcobatus Flat, with which it is connected by a thin ribbon of Recent gravels, and trends southeast to Ash Meadows, and thence south 60 miles to a point where it swings around the Funeral Mountains to join Death Valley. Within the area mapped the valley is about 10 miles wide at the north end, contracts in the middle, and expands to a width of 21 miles at the southern border of the area. The valley sends branches well into the surrounding mountains. From the summit south of Currie Well the valley descends 1,600 feet in 10 miles and but 500 feet in the next 20 miles. The cross section of the valley is comparatively flat except in the northern part. The lowest portion within the area shown on the map lies at an elevation of 2,500 feet. The channels of the so-called Amargosa River and of Fortymile Canyon, each characterized by bowlders and coarse gravel, are slightly depressed beneath the surface of the valley. The valley is without grass, but is covered with a sparse growth of creosote bush and other shrubbery, while mesquite trees grow north and east of Ash Meadows.

The Recent detrital deposits in the southern portion of the valley are fine grained, and the wind has heaped these into dunes. The most important group of sand dunes, 3 miles south of Rose's Well, is shown on the map. The Amargosa mine at Rhyolite has been sunk 330 feet in Recent gravel, and in some portions of the valley the Recent alluvial deposits are probably 500 or more feet deep. Inliers of Tertiary and Paleozoic rocks are, however, common. Many hills and ridges of Paleozoic limestone protrude from the gravel in the southeast corner of the area. The line of Paleozoic inliers northwest of the sand dunes is probably the crest of a ridge formerly connected with the Amargosa Range, but now separated from it by a mantle of detrital material.

The springs at Ash Meadows have already been described (p. 20). Water has been obtained at the following depths in three wells in the valley: Rose's Well, 208 feet; Miller Well No. 1, 183 feet; Miller Well No. 2, 70 feet. The water in all cases is good, although that of Rose's Well has a slight saline taste. The log of Miller Well No. 1, as reported by Mr. W. F. Miller, is as follows:

Log of Miller Well No. 1, Amargosa Desert.

	Feet.
Gravel	2
" Cement," indurated gravels	70
Gravel	4–6
" Cement "	10–12
Sand	6–7
" Cement "	78
Gravel (water stratum)	11

Playa and alluvial slope deposits of the older alluvium outcrop in a number of places in the Amargosa Desert, and these beds probably underlie at no great distance the south end of the valley. The hills in the desert 8 miles south of Bullfrog are composed of angular and subangular bowlders and sand derived from the surrounding hills, and among other rocks basalt occurs. The material closely resembles that of the present alluvial slopes. It has been eroded into hills, and its bowlders are somewhat weathered. It is without doubt an alluvial slope contemporaneous with the playa deposits next described. South of Rose's Well exposures of ancient playa deposits are common, and the northern border of the old playa appears to have been north of this well. At the southern border of the area surveyed this playa was probably 20 miles wide. The deposits now occur as low, white domes and dissected mesas, which are particularly abundant southeast and east of Miller Well No. 2. Six miles west of the Big Dune the deposits consist of fine-grained clay, in which are embedded many grotesque lime-carbonate concretions. Bedding planes are rare, but appear to be horizontal. The small deposit 1 mile north of Rose's Well is similar except that a few small, well-rounded pebbles occur in the clay and that a 1-foot layer of white porous limestone is interbedded with it. These beds at Ash Meadows form white limestone-capped mesas and buttes, which are landmarks for miles. A section of the bluff north of Ash Meadows is as follows:

Section of bluff north of Ash Meadows.

Feet.

1. White, carious, dense limestone, with many lenticular openings parallel to the bedding. The limestone is finely banded parallel to these openings and has the characteristics of chemically precipitated limestone_____ 10

2. White, fine-grained clay, massive, inclosing small crystal aggregates of selenite_____ 40

3. Limestone like No. 1_____ 1

4. Clay like No. 2_____ 20

5. Limestone like No. 1_____ ½

6. Clay like No. 2_____ 30

These beds lie horizontal, although Campbell [a] states that farther south they dip to the west. The playa deposits and gravels are without doubt members of the older alluvium, which is considered in part of late Pliocene and in part of Pleistocene age.

AMARGOSA MOUNTAIN SYSTEM.

The Amargosa Mountain system, including the Amargosa Range, Gold Mountain, Slate Ridge, and the Bullfrog Hills, is the most important uplift in the area under discussion. The mountain mass, which has a northwesterly trend, lies between Death Valley on the west and Sarcobatus Flat and the Amargosa Desert on the east. Geologically this mountain system is characterized by the widespread distribution of early Paleozoic rocks and in its northern portion by the presence of granite.

AMARGOSA RANGE.

TOPOGRAPHY AND GEOGRAPHY.

The mountain range between Sarcobatus Flat and the Amargosa Desert on the east and Death Valley on the west is designated the Amargosa Range by the United States Geographic Board, which also calls that portion of the range north of Boundary Canyon the Grapevine Mountains and that south of the canyon the Funeral Mountains. The mountain range as a whole is a unit, the most natural division on physiographic and geologic grounds being at Furnace Creek, south of the area surveyed, as suggested by Spurr.[b] The range here changes its trend; moreover, to the south of Furnace Creek it is composed principally of Tertiary sedimentary rocks and to the north of much older rocks.

The sinuous crest line of the Amargosa Range trends N. 40° W. and roughly coincides with the California-Nevada boundary line.

---

[a] Campbell, M. R., Bull. U. S. Geol. Survey No. 200, 1902, p. 14.

[b] Spurr, J. E., Bull. U. S. Geol. Survey No. 208, 1903, p. 187.

The northeastern side of the range is not impressive, the crest line rising only 2,000 to 4,000 feet above the Amargosa Desert and Sarcobatus Flat. On the other hand, the crest is from 5,000 to 7,000 feet above Death Valley, and in consequence the southwestern slope is steep, deeply cut by canyons, and set with numerous rugged peaks and pinnacles. The ridge culminates near its north end in Grapevine Peak, 8,705 feet above sea level. Wahguyhe Peak, with an elevation of 8,590 feet, is a magnificent truncated cone.

The Amargosa Range from Death Valley presents a somber appearance, the grays and blacks of the Paleozoic rocks predominating. The northern and middle portions of the range, as seen from the Amargosa Desert and Sarcobatus Flat, are colored by the rather brilliant reds, yellows, white, and grays of rhyolite flows.

The valleys heading in the range from the east are wide and gently sloping, and in some places, as at Indian Pass, broad detritus-filled embayments extend almost to the range crest. The stream channels on the southwestern side of the range are deep canyons, which, near the higher peaks, are sunk 3,000 feet beneath the sharp ridges on either side. The tortuous stream bed of Titus Canyon is only 40 feet wide, and from it the rocks rise sheer 200 feet, while the slope to the ridges 2,000 feet above is very steep. The canyon walls to a height of 40 feet are water polished, and cloudbursts have recently lodged bowlders 50 feet above the stream bed. The smoother reaches of the stream gradient are separated by transverse rock walls from 10 to 100 feet high. These barriers, which in most regions would cause waterfalls, are formed of resistant rock strata.

The eastern side of the range, from Chloride Cliff southward to the boundary of the area here discussed, is topographically older than the rest of the range, being a land surface to be correlated with similar tracts in the Panamint, Belted, and Kawich ranges. (See pp. 99, 119, 202.) A smaller area of this older topography is situated north of Thimble Peak. These portions of the range are characterized by dome-like mountains, shallow and gently sloping valleys, deep soil, and comparatively few and inconspicuous rock exposures. The area north of Thimble Peak is surrounded by deep-cut canyons, and the drainage lines pass precipitously from the mature upland to the young gulches. This mature topographic surface at one time extended over the whole mountain range, but is now preserved only where advantageously situated. Since it forms the eastern slope of the range south of Chloride Cliff and is not present on the western slope, it may be inferred that the grade of the streams on the west has been greatly increased by a rather recent depression of Death Valley. This depression occurred probably in early Pleistocene time, since a similar surface in the Panamint Range was developed before the outflow of late Pliocene or early Pleistocene basalt. (See p. 202.)

The crest of the Amargosa Range above 6,000 feet, at a distance of 8 miles northwest and southeast of Grapevine Peak, is partly covered by a sparse growth of piñon and juniper. The largest trees are 1½ feet through and from 20 to 25 feet high. The Paleozoic rocks to the west of the crest line are bare even above an elevation of 6,000 feet. Some grass grows in the higher valleys on the northeastern slopes of the range. South of Chloride Cliff the Grapevine Mountains have but few springs, and the largest of these is said to be poisonous and is shunned by the Indians. North of Chloride Cliff springs are comparatively abundant. The largest of these, Keane Spring, flows about 1,350 gallons per day; the two Willow springs and Daylight Spring, from 500 to 1,000 gallons each. The other springs are smaller, Hole in the Rock, when full, containing only about 10 gallons.

GENERAL GEOLOGY.

The formations exposed in the Amargosa Range, from the oldest to the youngest, are as follows: Prospect Mountain quartzite (?), Pogonip limestone, Eureka quartzite, Lone Mountain limestone, post-Jurassic granite, pre-Tertiary diorite porphyry, biotite andesite, earlier rhyolite and contemporaneous basalt, Siebert lake beds, later rhyolite and biotite latite, and basalt.

SEDIMENTARY ROCKS.

Prospect Mountain quartzite (?).—The Amargosa Range south of a line somewhat north of Chloride Cliff is formed of a series of quartzite beds with intercalated schists and marble. The large hill group north-northeast of Lee's Camp and the small boss-like hills northwest of the Big Dune are of the same formation. This series, which includes several thousand feet of sedimentary rocks, is considerably metamorphosed, the alteration being more evident in the schists and the quartzites than in the marble.

The quartzites are usually white, pink, or gray in color, although on the one hand a cement of specular hematite and on the other much clayey matter in the quartzite change the color to black. The normal, rather pure quartzose rock is fine to medium grained, and is in many places thin bedded. Cross-bedding and ripple marks are locally present. The quartzite passes into conglomerate with well-rounded pebbles, which reach a maximum diameter of one-half inch. The pebbles are of quartz, of vein or pegmatitic origin, white quartzite, black jasperoid, and schist. The development of sericite, particularly in the conglomeratic facies of the quartzite, imparts a greenish tinge to the rock. On microscopic examination this sericite proves to be contemporaneous with the quartz cement, which secondarily enlarges the clastic grains of quartz. Mashing, besides developing sericite, has locally elongated the pebbles of the conglomerate, forming a schistose conglomerate.

The more impure quartzite grades into slaty shales and schists. The shales are fine grained and of a greenish or dark-gray color, with laminæ from one-sixteenth to one-half inch thick. The slaty shale has usually a few muscovite plates on its parting planes and passes, with gradations, into fine-grained bluish or greenish-gray sericite schists. The schistosity is usually parallel to the original bedding of the rock, but in places cuts it at all angles. In the schist deep-red garnets, small tabular hexagonal greenish plates (ottrelite), greenish-gray micaceous aggregates (chlorite), or grayish prisms (andalusite) are developed. Under the microscope these schists appear without original structures. Fine shreds of sericite, with chlorite, in some thin sections are well aligned parallel to the schistosity, and between such bands are fine mosaics of quartz and a little orthoclase. Among the phenocrystic minerals of the series, clearly of later origin than the groundmass, are staurolite, andalusite, ottrelite, biotite, chlorite, garnet, ilmenite, and magnetite. Zircon, graphite, and rutile are present locally. Small crystals of tourmaline, cutting the schistosity in all directions, are present in most of the thin sections. The occurrence of this mineral 10 miles or more from the nearest granitic outcrop possibly indicates the presence of a considerable body of granite beneath the Amargosa Range. The interbedded marbles are fine- to medium-grained rather dense rocks, of buff, pink, purple, gray, or white color. Much of the marble is finely laminated. The presence of clastic grains and small pebbles of quartz and of cross-bedding indicates that the original limestone was deposited in shallow water. Through metamorphism not only has the rock been thoroughly recrystallized and locally somewhat silicified, but muscovite has been developed on the parting planes, and calcite veins, cutting it in all directions, have been formed.

Interbedded with and sending short arms out into the schist at Poison Spring are thin discontinuous sheets of hornblende gneiss, composed of a fine mosaic of quartz and feldspar, with much fibrous greenish-black hornblende and in certain bands red garnet. Under the microscope the mosaic is seen to be composed of quartz, orthoclase, and plagioclase. In the mosaic are large skeletons of hornblende, garnet, zoisite, and titanite. Apatite and magnetite are present as accessory minerals. Chlorite is secondary to hornblende, sericite, or kaolin, and calcite to the feldspars, and zoisite to the mutual alteration product of hornblende and plagioclase. A sheet of hornblende schist cuts across the bedding planes of marble and schist at a low angle 200 yards west of the Chloride Cliff camp, near the head of a gulch. The sheet is over 200 feet thick and is at least one-fourth mile long. This schistose rock is composed largely of hornblende, with some quartz and feldspar. Red garnets which reach a maximum diameter of one-half inch are abundant in some portions of it and are also present in the adjacent sericite schist. The hornblende schist is composed of the following minerals, named in the order of their abundance: Hornblende, quartz, biotite, plagioclase (oligoclase in part), orthoclase, garnet, magnetite, and apatite. The rock has the form of a sheet injected in the Cambrian rocks. This schist and the gneiss above described, each of which is probably a mashed basic igneous rock, are the oldest igneous rocks known in the area surveyed. The hornblende rocks are older than the post-Jurassic granite, which is everywhere massive, and since they have suffered approximately the same metamorphism as the rocks which they inject they are probably of Paleozoic age.

The sedimentary rocks of this series were laid down in a shallow sea in which the conditions of deposition were constantly and rapidly changing. Fragmental material was carried into the sea even during the deposition of limestone, since clastic quartz grains are common in it.

The series is identical with that of Tucki Mountain, in the Panamint Range. Its age was not definitely determined, since unfortunately the contact with the Ordovician and Silurian rocks lies between two traverses made across the range. Lithologically it more closely resembles the Prospect Mountain quartzite (Cambrian) of the Specter Range than any other formation. The two are both quartzites, with shale and conglomeratic facies containing similar pebbles. Limestone was, however, not noted in the quartzite of the Specter Range. While tentatively considered Cambrian, these sedimentary rocks may in reality be of pre-Cambrian age. The severe metamorphism of the series indicates that the rocks have undergone folding not suffered by the Ordovician rocks farther north, and that they are hence of pre-Ordovician age.

Pogonip limestone.—The Pogonip limestone extends from Keane Springs northwestward to Grapevine Canyon. It also covers a large area north of Grapevine Springs and smaller areas in the vicinity of and north of the Staininger ranch, east of Cave Rock Spring, and 4 miles east of boundary post No. 94. In the large area north of Keane Spring from 2,000 to 3,000 feet of limestone is exposed. It is a fine- to medium-grained dense rock, most of it dark gray or black in color, though some beds are light gray. The laminæ are as a rule from one-sixteenth to one-fourth inch thick, while the bedding planes are from 4 inches to 40 feet apart, heavy bedding being rather characteristic. Calcite veinlets anastomose throughout the limestone mass, and in Titus Canyon areas of coarse white calcite blotch the limestone. A thin bed of white quartzite is interbedded with the limestone in the lower part of the series at Cave Rock Spring. Similar quartzite and black flint occur in the limestone north of Grapevine Springs.

The Pogonip limestone is rather fossiliferous, the fossils being of Ordovician age. From the limestone north of Grapevine Springs poorly preserved fossils were collected, which Mr. E. O. Ulrich identified as *Orthoceras olorus* and *O. perroti*, other *Orthoceras* fragments, and a fragment of a pelecypod probably of the genus *Clionychia*. These fossils indicate a middle Ordovician horizon. Mr. F. B. Weeks [a] collected Ordovician fossils from the main Paleozoic area, 5 miles southeast of the Staininger ranch, and Mr. G. K. Gilbert [b] collected fossils in Boundary Canyon which he considered early Silurian or Cambrian.

Eureka quartzite.—Overlying the Pogonip limestone near Boundary Canyon is about 800 feet of quartzite. It is well exposed at Daylight Spring and is a pink, rather pure quartz rock of medium grain. The bedding planes, which are rather massive, are emphasized by many fine laminæ varying slightly in color and in size of the constituent grains. Conglomeratic bands containing well-rounded pebbles one-half inch in diameter occur. Some of the pebbles are of quartzite, possibly derived from Cambrian rocks. Thin layers of black fine-grained argillaceous quartzite and of olive-green or brown slaty shales are also interbedded with the normal quartzite. This rock is correlated, with considerable confidence, with the Eureka quartzite (Ordovican).

Lone Mountain limestone.—Overlying the supposed Eureka quartzite in the vicinity of Keane, Daylight, and Willow springs is about 300 feet of limestone which resembles the Pogonip closely, although less distinctly bedded. This is probably the Lone Mountain limestone [c] of the Eureka section. The inlier 4 miles north of the Staininger ranch is composed of fine-grained dark-gray limestone, with interbedded black flint, white or pink quartzite, and limy shales. Imperfect silicified fossils, one of which Mr. E. O. Ulrich considers a late Silurian coral (genus *Cladopora*), were collected by the writer from this locality. Other small inliers in this vicinity may in realty be the Lone Mountain limestone rather than the Pogonip.

Siebert lake beds and contemporaneous deposits.—The rather large mass of later rhyolite and biotite latite to the southeast of Staininger's Ranch has between its flows layers of white or light-colored in-

[a] Spurr, J. E., Bull. U. S. Geol. Survey No. 208, 1903, p. 188.
[b] Gilbert, G. K., U. S. Geog. Survey W. 100th Mer., vol. 3, 1875, pp. 34, 169, 181.
[c] Hague, Arnold, Mon. U. S. Geol. Survey, vol. 20, 1892, pp. 57–59.

coherent tuffaceous sandstone, which readily breaks down into sand. The section at this point is as follows:

Section southeast of Stainingers Ranch.

	Feet
Lava, largely biotite latite	375
Rhyolite flows and some tuffaceous sandstone	100
Tuffaceous sandstone and thin rhyolite flows	375
Rhyolite flows	75
Tuffaceous sandstone	225

Between the lava and the sediments are minor erosional gaps, but these are probably current erosional unconformities, and the series is approximately all of one age. Bedding planes from 2 inches to 20 feet apart are well developed at some places.

Locally the tuffaceous sandstone lies unconformably upon folded Pogonip limestone. It is closely similar to that of the Mount Jackson hills and, like it, is overlain by siliceous lavas. This formation is, without much doubt, the Siebert lake beds.

A number of small areas of conglomerate occur on the eastern slope of the Amargosa Range, east of boundary post No. 94, southeast of Daylight Spring and north of Willow Spring. The rocks, varying from loosely to rather strongly consolidated material, consist of conglomerates and arkoses with a little shale. They range in color from green to red, yellow, or brown. Bedding planes, which are from 1 inch to 2 feet apart, are well developed. The pebbles in the conglomerate, which are very well rounded and many of them highly polished, reach a maximum diameter of 4 inches. They are derived principally from the Paleozoic rocks near by, with here and there a pebble of fine- to medium-grained granite, and in the area east of boundary post No. 94 some of earlier rhyolite. These rocks appear to be the shore deposits of a lake in which waves swept with considerable force. Material was transported from considerable distances, since granite does not now outcrop in the vicinity. Since the conglomerate contains pebbles of rhyolite and none of basalt, the deposits are younger than the rhyolite (early Miocene) and older than the basalt (Pliocene-Pleistocene), and in consequence of middle or late Tertiary age, presumably Miocene. They are approximately contemporaneous with the lake in which the Siebert lake beds were deposited, and it is believed that they are shore deposits of that lake. The conglomeratic lake beds of the Funeral Mountains, south of the area mapped, are to be correlated with these conglomerates.

Older alluvium.—From a point 8 miles north of Surveyors Well to a point 3 miles north of Mesquite Spring the western flank of the Amargosa Range is formed of older alluvium, and the same formation covers a considerable area west of boundary post No. 85. This deposit consists of clays and bowlder beds similar to those of Death

Valley and the Panamint Range and, like them, is considered in part of Pliocene and in part of Pleistocene age.

Probably representing stream gravels contemporaneous with the older alluvium are small areas of conglomerates in Boundary Canyon and in the gulch in which Poison Spring is situated. In Boundary Canyon, near Hole in the Rock Spring, reddish conglomerate lies upon the eroded surface of the Paleozoic rocks, from which the bowlders have been derived. These bowlders reach a maximum diameter of 3 feet. The stream bed of the Poison Spring gulch is largely cut in calcite-cemented conglomerates. At the mouth of the gulch these beds are 40 feet above the stream, but 2 miles above Poison Spring they are at stream level, indicating that the present grade is steeper than that of the stream which deposited these gravels. Calcite veins, many of which are banded, fill joint fissures in the conglomerates. In a few places narrow seams of limonite indicate that some pyrite was deposited with the calcite.

IGNEOUS ROCKS.

In addition to the hornblende gneiss and schist already described (p. 163), the igneous rocks of the Amargosa Range include post-Jurassic granite, pre-Tertiary diorite porphyry, and several Tertiary lavas.

Granite porphyry.—Dikes of granite porphyry cut the Pogonip limestone 6 miles northwest of the Staininger ranch. The rock has a grayish-pink fine-grained groundmass, which is exceeded in bulk by the phenocrysts. Unstriated pinkish-gray feldspars, which reach a maximum length of one-fourth inch, are more abundant than small rounded quartz phenocrysts. Rarely a little chlorite stained by iron suggests the former presence of biotite. Under the microscope the groundmass appears as a microgranitic aggregate of quartz, orthoclase, and a little plagioclase. With the phenocrysts of orthoclase and quartz are some of biotite altered to chlorite and calcite and a few of plagioclase. This granite porphyry is probably one of the post-Jurassic granites.

Diorite porphyry.—Fragments of a greenish-gray much-altered igneous rock occur in the desert gravels on the eastern slope of the range west of Rose's Well, and dikes of this rock probably cut the Cambrian sediments. The rock is apparently the pre-Tertiary diorite porphyry.

Biotite andesite.—The oldest of the Tertiary lavas is a biotite andesite, which occurs in a number of small areas between Grapevine Canyon and Willow Spring, being particularly abundant near Mexican Camp. The biotite andesite is a dense lithoidal or semiglassy rock, which is gray, green, brick red, or purplish red, the last color evidently being due in many cases to alteration. Phenocrysts are usually somewhat subordinate in bulk to the groundmass. Biotite and lath-shaped feldspars, many of them considerably altered, are as a rule equally abundant. The feldspar phenocrysts are slightly larger than those of biotite, but they are nowhere over one-fourth inch long and are in most places much smaller. Flow lines are present in the groundmass and many of the phenocrysts show flow orientation. Much of the rock is vesicular. The vesicles reach a maximum length of $1\frac{1}{2}$ inches and are more or less completely filled by chalcedony and chlorite. Hyalite, chalcedony, and quartz occur along joint planes.

One thin section shows under the microscope as a biotite andesite with a groundmass composed of plagioclase laths in a rather ill-defined gently birefringent substance. The phenocrysts include plagioclase, biotite rimmed by beads of magnitite, orthoclase, and magnetite. The small plagioclase phenocrysts are simple and the large ones complex aggregates. Zircon is a rare accessory mineral. Another thin section, although clearly from the same magma, is somewhat more acidic, having affinities with latite. Quartz and orthoclase are present in the microfelsitic groundmass and orthoclase phenocrysts are as abundant as those of plagioclase.

The biotite andesite occurs in flows which were eroded prior to the outflow of the earlier rhyolite. On the range crest 4 miles south of Wahguyhe Peak the rhyolite contains small fragments of andesite. The andesite is in consequence younger than rhyolite and is probably of late Eocene age. The more acidic facies of the rock are mineralogically rather like the monzonite porphyry of the Kawich Range, which is older than the earlier rhyolite of that range, but which is probably contemporaneous with the earlier rhyolite of the Amargosa Range. The andesite and monzonite porphyry are probably, in a broad way, contemporaneous.

Earlier rhyolite.—Rhyolite forms the crest and eastern slope of the Amargosa Range from Thorp's mill to a point 2 miles northwest of Willow Spring, and thence to a point $2\frac{1}{2}$ miles southeast of Daylight Spring it forms the eastern slope of the range. East-southeast of this point several small rhyolite areas occur.

The rhyolite series comprises a considerable number of siliceous volcanic rocks, which lie in well-defined flow beds. The predominant facies is a lilac, gray, white, or pink rhyolite, with lithoidal groundmass and medium-sized phenocrysts of equal volume. Of the phenocrysts, colorless or slightly smoky quartz and glassy, unstriated feldspar are, as a rule, equally abundant, although the feldspar may predominate. Biotite phenocrysts are usually sparse and of smaller size. Other facies are gray glasses without phenocrysts, but with highly developed perlitic parting, and black glass, with feldspar and fewer quartz phenocrysts. Similar rocks from other ranges prove, on microscopic examination, to be latite, and the series probably includes latite and siliceous dacite in addition to rhyolite. Compact aphanitic red facies of rhyolite are also common, and in these flow banding, lithophysæ, and spherulites are well developed. The rhyolite of Shale Peak (elevation, 7,370 feet), 4 miles southeast of Wahguyhe Peak, is a faded brick-red rock banded with pink lines. Exfoliation breaks this rock, which is rather poor in phenocrysts, along the flow lines into numberless platy fragments, which caused the members of a United States Coast and Geodetic Survey party to give the peak its expressive name. Many angular inclusions of rhyolite of slightly different character occur in the lava, forming flow breccias. The rhyolite is in some places kaolinized and in others silicified. (See p. 47.) It is probable that much of the rhyolite that is now, through kaolinization, white or light gray in color, was formerly brick-red, since at certain places bleaching has occurred along joint fractures and irregular cracks, forming mottled fantastic patterns in reds and whites.

Beds and lenses of red, white, and greenish-white incoherent sandy rhyolitic tuffs, containing feldspar, quartz, and biotite crystals, are interbedded with the lavas. In these tuffs pebbles of rhyolite, which in rare instances reach a diameter of 1 foot, are abundant. All gradations exist between the normal igneous rock and these sediments deposited in small lakes or subaerial basins.

The earlier rhyolite of the Amargosa Range occurs in flows from 10 to 300 feet thick, superimposed one upon another, which impart to the eastern side of the range its varicolored banding. The thickness of the series is unknown, although it reaches at least 3,000 feet. The greater portion outflowed as lava, but at various times during the extrusion rhyolitic material was deposited in local sheets of water and at others dust from explosive eruptions formed subaerial accumulations. The vents from which the rhyolite flowed are probably situated near the crest line, but their position was not determined in the course of the present work. Where the rhyolite flows vary considerably in hardness the more resistant beds form benches and the softer beds are covered by talus, a method of weathering well developed on the eastern side of the north end of the range. Where vertical systems of joints are prominently developed, outcrops are bounded by vertical planes and rock pinnacles form. If the rhyolite flows are of approximately equal hardness and joints are not prominent, rounded bosses and smooth domelike hills result. Such forms are well seen near Shale Peak. Perhaps the most common topographic forms are cones, a mean between the pinnacles of the well-jointed rhyolite and the domelike hills of the more massive facies.

The earlier rhyolite lies upon the eroded surface of the Paleozoic rocks and the biotite andesite and near contacts contains numerous pebbles of these rocks. The later basalt lies upon the eroded surface of the rhyolite. This rhyolite is practically continuous with that of the Bullfrog Hills and is of similar lithologic character. It is roughly to be correlated with the earlier rhyolite of the Kawich, Belted, and Reveille ranges, probably of early Miocene age.

Earlier basalt.—Basalt fragments are locally contained in the rhyolite flow breccias, and at several places thin lenses of basalt, too small to show on the map, are interbedded with the rhyolite. The best example is at the Happy Hooligan mine, 12 miles south of west of Rhyolite. Mr. F. L. Ransome [a] states that recent mining development has here exposed a basalt bed from 25 to 50 feet thick between the rhyolite and the Pogonip limestone. The rhyolite of the Bullfrog Hills and the Kawich Range also contains minor flows of basalt, a fact further indicating the contemporaneity of the two rocks.

Later rhyolite and biotite latite.—To the southeast of the Staininger ranch is an area covering about 12 square miles of rhyolites and latites interbedded with the Siebert lake beds. The rhyolites are the older of these lavas and the latites the younger.

The rhyolites are semitransparent to opaque, glassy or semiglassy rocks, with few quartz, feldspar, and biotite phenocrysts. In color they range from light gray, through red, to black. Flow breccias are common. Perlitic parting, which causes the rock to disintegrate into pebble-like masses, is well developed in many of the flows. Flow folding, with overturned isoclinals, in places with horizontal axes, is not unusual. The spherulites, which are very abundant in some flows, making up at least one-third of the rock, are similar to those already described from the Mount Jackson hills (p. 67). In some cases large spherulites, rudely globular and from 3 inches to 1 foot in diameter, inclose smaller ones. The large spherulites are cut by horizontal flow lines, while the smaller have the typical radiate structure. Lithophysæ also occur.

The latites are white semipumiceous rocks with myriad bronze-brown hexagonal plates of mica and less conspicuous phenocrysts of feldspar. In some instances the biotite plates have an excellent orientation in the place of flow and the rock breaks into platy masses along the flow banding. A single thin section examined under the microscope showed the rock to be a biotite latite, with groundmass of a vesicular colorless glass. The phenocrysts exceed the groundmass in bulk, and orthoclase and biotite are more abundant than plagioclase. Apatite is present as an accessory mineral.

The later rhyolite is similar lithologically to the rhyolite associated with the Siebert lake beds of the Mount Jackson hills, and the two series are probably approximately contemporaneous. In the

---

[a] Oral communication.

upper portion of the series a few basalt flows are apparently inter-bedded with the rhyolite. At another place the basalt clearly cuts the Siebert lake beds and the rhyolite. (See p. 171.) This shows that while toward the end of the period of rhyolite extrusion some basalt probably outflowed, for the most part the basalt is younger. The rhyolite and latite are probably of late Miocene and early Plio-cene age.

Later basalt.—Basalt is the youngest of the igneous rocks. Its flows cover a considerable area at the north end of the range, and a number of smaller areas occur on either side of Grapevine Canyon. A dike of basalt 4 feet thick cuts the younger rhyolite and Siebert lake beds 2 miles north of the Staininger ranch. This was probably one of the vents through which the lava outflowed, while a small, red, conical hill to the south of the Staininger ranch–Thorp road, 5 miles from the ranch, appears from a distance to be a volcanic cone. The two small buttes in Sarcobatus Flat, north and northeast of Currie Well, seen only from a distance, are probably basalt. Remnants of flows cap rhyolite hills southeast of Daylight Spring, and at one locality a dike of basalt cuts the Pogonip limestone. This dike may be the ancient vent from which the basalt in this vicinity flowed. The basalt of the dike is reddish brown in color and much altered, but on microscopic examina-tion appears to have been originally an olivine basalt. The basalt from the flows in the vicinity is black, vesicular, and without promi-nent phenocrysts. The basalt of Grapevine Canyon is for the most part black and dense, although basal portions of the flows (in many places flow breccias) are vesicular and red. The more vesicular rocks are locally covered by a layer of basalt banded horizontally by flow lines, and this in turn by more massive beds which show sphe-roidal weathering. Phenocrysts, while usually small, are abundant in many outcrops and include glassy laths of plagioclase and rounded grains of olivine, either fresh and glassy or iron stained. Granular calcite, probably derived from the alteration of the basalt, fills vesi-cles in the basalt and veins cutting it. A thin section cut from basalt from the low mesas 4 miles east-northeast of the Staininger ranch proves to be a quartz-bearing olivine basalt with holocrystalline groundmass. The quartz is deeply embayed and surrounded by a corona of augite columns of the groundmass.

The basalt from the south end of the Amargosa Range overlies the Siebert lake-bed conglomerates, and from its lithologic char-acter is considered to be one of the Pliocene-Pleistocene basalts. The basalt in the vicinity of Grapevine Canyon is probably in part con-temporaneous with the upper part of the later rhyolite and Siebert lake beds, but is for the most part younger and of late Pliocene or early Pleistocene age.

The Amargosa Range has been subjected to two main periods of folding and uplift, the earlier in pre-Tertiary and presumably post-Jurassic time, the later probably in late Miocene time, after the deposition of the Siebert lake beds. In the northern part of the range this later folding was somewhat earlier than in the southern part and appears to have succeeded the outflow of the earlier rhyolite and preceded the deposition of the Siebert lake beds. North of Wahguyhe Peak the Paleozoic rocks and the earlier rhyolite are so folded together that the differentiation of the two periods of folding is difficult. Even here, however, the Paleozoic rocks are more profoundly folded. The earlier folding was accompanied by reversed faulting and probably in the last stages by normal faulting; the later folding was accompanied by normal faulting.

The major direction of the early folds was parallel to the trend of the range, but accompanying them are cross folds, which west of Grapevine Peak equal in importance and south of Indian Pass predominate over the northwest-southeast folds. In the vicinity of Grapevine Peak these cross folds were intensified by later folding. In the central portion of the range there is an anticlinorium with northwest-southeast axis, composed of two main anticlines with a syncline between. The syncline at Keane Spring is comparatively narrow; at Boundary Canyon it is broader and more open; and in the canyon in which Tule Spring is situated it is represented by two minor synclines and an intervening anticline. At Chloride Cliff the syncline is broken by a normal fault. Many of the minor folds are closely appressed isoclinals or fan folds, which in some instances have horizontal axes. The passage from isoclinal folds to reverse faults is by no means unusual. South of Indian Pass the structure is that of an anticlinorium with northeast-southwest axis, of which the long northwest limb dips rather gently to the southeast, while the shorter southeast limb dips somewhat more steeply to the northwest. Superimposed upon the main folds are minor folds, many of which are rather steep. The Pogonip limestone north of Grapevine Springs is folded into a gentle arch with east-west axis.

Before the later folding was inaugurated the range was worn down to an inconspicuous ridge, and the eastern slope and the whole range south of the area mapped formed the bottom of the Miocene lake. It is to the later folding and succeeding minor uplifts that the range owes its height. South of Daylight Spring the rhyolites dip at angles of 20° to 40° away from the range, while the Siebert lake conglomerates in this vicinity have a dip away from the range which in some instances reaches 70°. North of Daylight Spring the earlier rhyolite is folded into a synclinorium whose axis strikes N. 70° E., with its trough situated about 4 miles north of Grapevine Peak. Normal strike faults having the same strike are common in the vicinity of Thorp's mill and form numerous fault scarps. The Siebert lake beds south of the Staininger ranch have been flexed and are broken by small normal faults of north-south strike.

Unimportant folding and faulting are recorded in the basalt and older alluvium areas. The basalt below the Staininger ranch dips toward Death Valley and has apparently been tilted westward along a north-south axis. Normal faults of 4 to 5 feet displacement, usually with north-south trend, cut the basalt in the vicinity of Grapevine Canyon. The older alluvium at Hole in the Rock Spring for the most part dips toward Death Valley, although northeasterly dips here and there indicate that since the deposition of the conglomerate the range has not only been uplifted, but also somewhat folded.

ECONOMIC GEOLOGY.

Active mining operations are being caried on at three points in the Amargosa Range within the area mapped—at the Happy Hooligan mine, at Chloride Cliff, and at Lee's Camp.

HAPPY HOOLIGAN MINE.

The Happy Hooligan mine is situated at the eastern base of the Grapevine Mountains, 9 miles northwest of Bullfrog. A number of test pits have been sunk at the contact of the eroded surface of the Pogonip limestone, with its interbedded quartzite and the Tertiary lava flows. The predominant lava is a kaolinized rhyolite, from which the feldspar phenocrysts have been largely removed by weathering. This is underlain by a flow of basalt which rests upon the limestone. At the contact with the limestone the basalt is reddened by hematite and altered to a greasy substance, which pans fine free gold, as does the limestone when decomposed. Gold values also occur where the lava comes in contact with the quartzite. The contact has been traced several hundred feet and values are reported from all the test pits. The ore is free milling, and no sulphides were observed. It is said to run $40 in gold per ton, the values appearing to be closely associated with limonite and hematite. Several springs are situated near the mine and fuel can be obtained within 6 miles. The road to Bullfrog is rather heavy.

CHLORIDE CLIFF.

Chloride Cliff is situated in Inyo County, Cal., on the crest of the Amargosa Range, 15 miles south of Bullfrog. At this point the Cambrian limestone strikes N. 15° W. and dips 30° E. A steeply inclined fault trending N. 50° W. cuts the limestone. The limestone along the fault is silicified and in it are disseminated galena and some chalcopyrite. The secondary ore is a porous yellowish lead carbonate with some malachite stains. The surface outcroppings are heavily stained by limonite. At another place a 4-foot vein of quartz lies between the gently dipping hornblende schist and the sericite schist and altered limestone above it. The quartz is free milling and heavily iron stained, and the crushed portions have been recemented by limonite. The writer traced the vein 200 feet, but it extends much farther. It is reported that one sampling across the vein gave values in gold averaging $1,500 per ton. At a third prospect a reversed fault cuts the schist member of the series, which here strikes N. 60° W. and dips 15° NE. A quartz vein 2 to 3 feet wide occupies the fissure. The quartz is stained by limonite, and pyrite is the only original sulphide seen. Small nodules of native copper are, in rare instances, associated with the limonite stains. The ore is said to assay from $13 to $500 per ton in free gold. The values are associated with limonite.

The prospects at Chloride Cliff are either fissure fillings or deposits along contacts. In some cases the quartz and accompanying sulphides probably filled open fissures; in others the limestone was metasomatically replaced. There is a distinct tendency for galena to be the predominant sulphide in limestone, and pyrite in the other country rocks. Since the deposition of the quartz the veins have been crushed and surface waters have produced from the original sulphides native gold, copper, cerussite, malachite, and limonite. The deposits are structurally similar to those which hold their values well with depth, and if surface enrichment has not been out of all proportion to the original sulphide values, further development is well justified.

Water is at present packed on burros from Keane Spring, 4 miles distant. Wood suitable for fuel occurs on the Amargosa Range, 10 miles north of the prospects, but at present must be teamed by way of Bullfrog. A fair road connects the prospects with Bullfrog, the shipping and supply point.

Ores from a number of mines in the Cambrian rocks in the vicinity of Chloride Cliff were examined, but the mines themselves were not visited. At the Keane Wonder mine a blanket vein of quartz, dipping gently westward, is said to lie along the bedding planes of the Cambrian rocks. The gold ore is quartz heavily stained by limonite and is said to be free-milling. The Trio mine, in the next gulch south of the Keane Wonder, is said also to be on a quartz vein in the Cambrian rocks. The ore is a limonite-stained quartz containing chalcopyrite and galena and some cerussite. Free gold is reported to occur in schist and limestone along a fault fissure down the gulch from Keane Spring. The ore of this deposit closely resembles that from the metasomatic replacement deposits of Chloride Cliff. The

Bismuth mine is situated on the road to Bullfrog, somewhat below the Chloride Cliff mine. The quartz, which is heavily stained by limonite, contains finely granular galena. The presence of some malachite probably indicates the former presence of chalcopyrite with the pyrite and galena.

LEE'S CAMP.

Lee's Camp is situated in the Cambrian rocks, 10 miles southwest of Rose's Well. The camp was not visited by the writer, but strong veins of quartz are said to exist there. The ore, which is said to mill free gold, is quartz and either dolomite or siderite, each heavily stained by limonite.

AREAS FAVORABLE FOR PROSPECTING.

Quartz veins are common in the metamorphosed Cambrian sedimentary rocks. They occupy joint and bedding planes and fault fissures, and in the latter case may contain angular fragments of the Cambrian rocks. In size they vary from mere stringers to veins 6 feet or more across, which can be traced for hundreds of yards. In the vicinity of Poison Spring these are particularly strong, and over restricted areas make up almost one-third of the rock. The quartz in most instances is white and semitransparent and appears barren. In others it has been faulted and brecciated and is stained by limonite, while pyrite and specular hematite are disseminated in it here and there. The mines of Echo Canyon, to the south of the area surveyed, are reported to be situated on similar quartz veins. Pyrite and hematite also occur in quartz veins on the small Cambrian hills to the northwest of the Big Dune, and these minerals and malachite are found in the quartz veins of the big Cambrian inlier to the southwest of the Big Dune.

In the Pogonip limestone of the central and northern portion of the Amargosa Range quartz veins are of minor importance. The veins, some of which contain pyrite and specular hematite, appear to be comparatively small. About 6 miles south of the Staininger ranch prospectors sunk a shaft 10 feet deep on a quartz vein stained with malachite. At this depth the vein divided into several stringers and the shaft was abandoned. The ore carried low gold and silver values. With the change of the country rock from the siliceous deposits of the Cambrian area to the calcareous rocks of the Silurian and Ordovician, calcite veins, probably in part contemporaneous with the quartz veins, become predominant. These are extremely abundant, but as a rule are less persistent and less mineralized than the quartz veins in the Cambrian. At Titus Canyon the calcite veins are of two ages. The older are formed of white crystalline calcite, much of which is very coarsely granular. The younger veins, of banded yellow or brown calcite, are still forming. River bowlders are here and there inclosed in these veins, and stalactitic calcite hangs from them in some canyon walls. The bands are either parallel to the walls or at right angles to them, while in certain places the borders of the veins are vertically banded and the centers horizontally banded. Many of the later veins fault the older.

Areas of silicified and kaolinized rhyolite in the Amargosa Range are shown on the map (fig. 4, p. 43). The rhyolite 8 miles southeast of Daylight Spring has been silicified, and small quartz veins cut it. The area has been located, but no information is at hand as to the value of the ore. Silicified rhyolite, with many small quartz veins, outcrops at several places east of Mexican Camp and 1 mile north of McDonald Spring. At the latter locality small quartz veins also occur in the biotite andesite. The quartz in this portion of the range is largely white or gray and compact. At Currie Well bowlders of silicified rhyolite, said to pan gold, occur as float, but the original position of the rock is not known. Silicified rhyolite stained by the yellow basic ferric-alkali sulphate (?) already mentioned (p. 50) covers a considerable area to the south of Grapevine Canyon. This rhyolite is doubtless worthy of careful prospecting.

BULLFROG HILLS.[a]

TOPOGRAPHY AND GEOGRAPHY.

The Bullfrog Hills lie west of Oasis Valley, between the Amargosa Desert and Sarcobatus Flat. They form a spur of the Amargosa Range and have an east-west crest line, from which an eroded mesa extends northward and rather sharp ridges stretch out to the south. The highest peak in the hills reaches an elevation of 6,035 feet. The Bullfrog Hills are bare of timber and are only sparsely covered with desert shrubs and cacti. The towns in these hills obtain water from Amargosa River (see p. 18) and from a number of springs in Oasis Valley and in the hills themselves. Among the latter may be mentioned Indian (30,000 to 40,000 gallons per day), Crystal, and Mud springs. Hicks Hot Springs, in Oasis Valley, have already been described (p. 20).

GENERAL GEOLOGY.

The formations exposed in the Bullfrog Hills include the following, the oldest being named first: Pre-Silurian schist, Eureka quartzite, Lone Mountain limestone, post-Jurassic pegmatite, pre-

---

[a] The geology and ore deposits of the Bullfrog district will be described in detail by Messrs. F. L. Ransome, W. H. Emmons, and G. H. Garrey in a forthcoming Professional Paper, and the geology of the area shown on the Bullfrog special sheet has been generalized from their map. The description here given refers more particularly to the area outside that covered by the special map.

Tertiary diorite porphyry, andesite, rhyolite, earlier basalt contemporaneous with rhyolite, Siebert lake beds(?), and later basalt.

SEDIMENTARY ROCKS.

Pre-Silurian schist.—Three miles west of Bullfrog is a low ridge of post-Jurassic pegmatite containing a vast number of schist inclusions. The rocks are present in practically equal amounts, and the area could as well be described as an area of schist intensely injected by pegmatite. The pegmatite, however, is more resistant to erosion and hence gives its white color to the low rounded ridge.

The schist varies from a finely foliated biotite schist to an " eye " schist, containing many lenses of pegmatitic feldspar from one-sixteenth inch to 2 inches long. These eyes are elongated parallel to the schistosity, the shortest diameter usually being one-fourth the length of the longest. Some of the schist is composed of minute scales of biotite of common orientation, while other portions also contain quartz, a little feldspar, and the large feldspar lenses already mentioned. The planes of schistosity dip at considerable angles, but are rarely minutely crenulated. A single thin section of the " eye " schist under the microscope proves to be a schist in which shreds of biotite and a little muscovite are arranged in folia. Between these is a fine-grained mosaic of orthoclase and quartz, with here and there a partial crystal of apatite or zircon. The pegmatitic " eyes," which are usually acidic plagioclase and in only a few cases quartz or orthoclase, are closely encircled by the biotite films.

The schists are presumably metamorphosed sedimentary rocks, but they may possibly be mashed igneous rocks. Mr. G. H. Garrey,[a] who examined the relations between the schist and the Pogonip limestone on the western border of the Bullfrog special area, states that the contact between the two rocks is not exposed, but that outcrops of limestone adjacent to the schist are comparatively unmetamorphosed and that the schist in consequence is presumably pre-Silurian. Lithologically this rock resembles Cambrian schists near Tokop, and although possibly of pre-Cambrian age it is tentatively correlated with the Cambrian.

Eureka quartzite.—A hillock rising from the rhyolite to the southeast of the limestone last described and a hill 2 miles south of Indian ranch are formed of quartzite interbedded with shale. Fragments of similar shale are included in Tertiary rhyolite 3½ miles northeast of Bullfrog. The Eureka quartzite of Bare Mountain extends across Amargosa River northwest of Gold Center. The quartzite of the hills first mentioned is gray, pink, or white in color, and of fine to medium grain. It appears to be a practically pure

---

[a] Oral communication.

quartzose rock, many portions of which are finely laminated. Associated with it are minor bands of brown, greenish, or purplish-gray shale, the laminæ of which are in places paper thin. The quartzite apparently underlies the limestone south of Indian ranch and lithologically resembles the Eureka quartzite (Ordovician).

Lone Mountain limestone.—Small bodies of limestone, in part, at least, of Silurian age, protrude from the Tertiary lavas and Recent gravels at four points—3 miles west of Bullfrog, three-fourths mile southwest of the Gold Bar mine, and 1½ miles south and 2½ miles southeast of Indian ranch. The limestone is dark gray, dense, and of medium grain. Finer grained facies closely resemble lithographic limestone. Interbedded with the normal type are layers of intraformational conglomerate. Thin seams of calcite locally cut the limestone in all directions. The limestone 1½ miles south of Indian ranch clearly overlies a quartzite which is probably the Eureka quartzite; if so this is the Lone Mountain limestone. Moreover, Mr. G. H. Garrey collected from the limestone inlier 4 miles northwest of Bullfrog a number of fossils which Mr. E. O. Ulrich states are of Silurian age.

Siebert lake beds.—One mile north of Crystal Spring is a small outcrop of an arkose conglomerate. The well-rounded pebbles, which reach a diameter of 1 foot, consist of quartzite, jasperoid, limestone, and medium-grained biotite granite. The first three are derived from Paleozoic rocks, while the granite resembles many post-Jurassic granites. The conglomerate is lithologically similar to the more conglomeratic of the Siebert lake beds of the Amargosa Range, and may well be contemporaneous with it. If such is the case the shore line of the lake must have been near Bullfrog. The relative ages of the rhyolite and the lake beds are unknown.

Older alluvium.—On the northeastern' face of the Bullfrog Hills and across Amargosa River to the east are hills and terraces dissected by the present streams. These hills and terraces are formed of detrital material similar to that now being deposited in stream beds and on the alluvial slopes. The deposits were made while Amargosa River was at a higher level and presumably are alluvial slope deposits of the older alluvium. Basalt bowlders (presumably of Pliocene-Pleistocene age) are contained in the upper portion of the beds.

IGNEOUS ROCKS.

Post-Jurassic pegmatite and alaskite.—Pegmatite with alaskite occurs with the pre-Silurian schist west of Bullfrog and dikes of pegmatite cut the Eureka quartzite near Gold Center. (See p. 155.) West of Bullfrog the pegmatite grades into alaskite, but the former rock is predominant. The most widely distributed facies of pegmatite is formed of white feldspar, less muscovite, still less smoky quartz, and a little pyrite. The largest feldspar individuals reach a diameter of 4 inches, while muscovite is in aggregates of plates from one-fourth to 1 inch across. Quartz veins of pegmatitic origin are also present and similar veins, probably of the same origin, cut the more feldspathic pegmatite. From this it is probable that a rock of granitic composition separated from the magma and at a later period a more quartzose rock was deposited from the residual liquid in fractures in the older pegmatite. Lens-like veins of practically pure white feldspar grade on the one hand into normal pegmatite, and on the other into the eyelike areas of the schist already mentioned.

The associated alaskite is a white, medium-grained rock composed of quartz and feldspar, with muscovite and a little altered biotite. The quartz and orthoclase grains are micropegmatitically intergrown along their borders and the biotite shreds are largely chloritized. Accessory minerals are not abundant, but include apatite, zircon, and magnetite, and apparently tourmaline. In the hand specimen the rock is considerably mashed and the microscope shows phenomena of granulation and recrystallization, including the deposition of muscovite, derived from feldspar, in fractures.

The pegmatite and alaskite inject the schist bed by bed, although in rare instances the planes of schistosity are crossed. The wider pegmatitic bands are 100 or more feet thick, while the thinner pegmatitic layers are but one-fourth inch wide, and many of them are separated only by an equal thickness of schist. The pegmatite cuts schists probably of pre-Silurian age and is older than the unmashed rhyolite of Tertiary age. On lithologic grounds its age is provisionally believed to be post-Jurassic and pre-Tertiary.

Pre-Tertiary diorite porphyry.—A few narrow dikes of diorite porphyry cut the pegmatite and schist 3 miles west of Bullfrog. The diorite porphyry is a dark grayish-green rock speckled with tiny white areas, and is evidently composed predominantly of a ferromagnesian mineral and feldspar. The rock has a medium-grained microgranitic groundmass of plagioclase, orthoclase, and some quartz. Columns of brown hornblende are very numerous, while larger phenocrysts of sericitized plagioclase are fewer. The accessory minerals, titanite, apatite, and magnetite, are very abundant. Epidote and chlorite are developed from hornblende or from hornblende and plagioclase. This rock is, without much doubt, to be correlated with the pre-Tertiary diorite porphyry widely distributed in the mountains on the western border of the area surveyed. (See p. 31.)

Biotite andesite.—A hill 2 miles northwest of Howell ranch and a small knob rising from the older alluvium three-fourths of a mile south of Indian ranch are of biotite andesite. This is a dense, flinty or lithoidal rock of medium, pinkish, or purplish gray color. Black glassy facies are less common. The small phenocrysts, feldspar laths, with slightly fewer black mica plates are subordinate to the groundmass in bulk. The feldspars are usually altered, but in some instances they are glassy and twinning striations are visible. From the bedded disposition of the various textural facies and the flow orientation, visible in many of the phenocrysts, the biotite andesite is probably a flow considerably eroded and masked by coverings of later formations. The groundmass under the microscope is seen to consist of a felt of plagioclase laths in a somewhat devitrified glass (hyalopilitic). Besides the simple laths of plagioclase, usually showing Carlsbad twinning, and the slightly bleached biotites, a rare orthoclase phenocryst is present, while certain ill-defined pseudomorphs, largely of serpentine, have the crystal outlines of hornblende. Ilmenite and apatite are present as accessory minerals.

On the west side of the hill northwest of Howell ranch much-altered pebbles and fragments of andesite are included in a tuffaceous member of the rhyolite series, and between the outflows of the two lavas there was apparently an erosion interval. The andesite is probably of late Eocene age, and is similar to and probably contemporaneous with the biotite andesite of the Amargosa Range.

Rhyolite.—The predominant formation of the Bullfrog Hills is a series of rhyolite flows. The most widely distributed facies is a lithoidal rhyolite of gray, pink, or purple color. The medium-sized phenocrysts, which usually exceed the groundmass in bulk, consist of glassy unstriated feldspar, slightly smoky quartz, and in many places biotite. Hornblende is also locally present. Less widely distributed types include aphanitic rocks with varicolored flow banding, resembling slaty shales; brown or greenish glasses, with or without phenocrysts, and gray glasses with perlitic parting well developed. Interbedded with the other members of the rhyolite series are rhyolitic sandstones and conglomerates, while Messrs. Ransome, Emmons, and Garrey found a thin limestone lens in the series. Through metamorphism, kaolinized and silicified facies arise. (See p. 47.)

The rhyolite is a series of igneous flows, with minor intercalated beds of sedimentary rocks, in part deposited under water and in part subaerially. Where the flows are thick the rhyolite in places weathers into bosses and low domes, but for the most part the varying resistances of the different flows to erosion give a bench and cliff character to the topography.

This rhyolite, which contains fragments of Paleozoic rocks and biotite andesite, on lithologic and structural grounds is correlated with the earlier rhyolite of the Amargosa and Kawich ranges. It is probably of early Miocene age.

Earlier basalt.—Interbedded with the rhyolite series in the Bull-frog district are several thin flows of basalt, and the latest of these is also found in the surrounding hills near the boundary of the special area. Dacite usually lies on the basalt and is evidently a later flow of a practically continuous period of extrusion. The basalt is a dark-gray or blue-black dense rock containing numerous laths of glassy striated feldspar and here and there a suggestion of an olivine or pyroxene phenocryst. Vesicular facies are common, and with the introduction of vesicles much of the rock takes on a red tinge. A single thin section examined proved to be an olivine basalt, with plagioclase, augite, and olivine phenocrysts in a groundmass composed of glass and tiny crystals of plagioclase, pyroxene, and iron ore (hyalopilitic).

Dacite is well exposed 1 mile southwest of Indian Spring. It is a purplish-gray rock, with dense groundmass and abundant phenocrysts, including glassy striated feldspar, black mica, and quartz. The feldspars very in size from one-sixteenth to 1 inch, the larger ones usually showing well-developed zonal growth. The quartz crystals reach a maximum diameter of one-fourth inch. Under the microscope the groundmass is seen to be a glass containing many tiny plagioclase laths (hyalopilitic). Besides the phenocrysts already mentioned, augite is present, while aggregates of heavily iron-stained serpentine may possibly be altered olivine.

Later basalt.—Large areas of later basalt cap the rhyolite in the northern portion of the hills and pass into northeastward-dipping mesas, which extend toward Pahute Mesa. The basalt, which is a dense black rock, locally vesicular, lies upon the eroded surface of the early Miocene rhyolite. It is similar lithologically to and appears to be an extension of the Pahute mesa basalt, and is, in consequence, probably of late Pliocene age.

STRUCTURE.

The isolated inliers of Ordovician-Silurian rocks are gently folded and somewhat faulted. Upon the major folds minor flexures, in some cases overturned isoclinals, are superimposed. The folding of the Paleozoic rocks probably occurred largely prior to Tertiary time, although the supposed Siebert lake beds appear to be conformable in dip with the Lone Mountain limestone. The rhyolite throughout the Bullfrog Hills is a mosaic intersected by numerous normal faults. These faults have been worked out in detail by Messrs. Emmons and Garrey in the area covered by the Bullfrog special map. The work carried on under the direction of Mr. F. L. Ransome shows that the predominant faults trend northeast, while a second set trends north and south. The later basalt is slightly tilted to the northeast.

The ore deposits of Bullfrog have recently been described by Ransome.[a] He shows that the ore occurs in and along fault fissures in silicified or kaolinized rhyolite. Silicification and kaolinization have been important processes of metamorphism throughout the Bullfrog Hills. The areas outside of the region shown on the Bullfrog special map so altered are delineated in fig. 4 (p. 43). Both the rhyolite and the andesite of the two domelike hills northwest of Howell ranch are silicified, and in addition seams of chalcedonic quartz fill solution cavities and fault and joint fractures. Other areas of silicified rhyolite are situated northwest of Crystal Spring, west of Indian Springs, and 4 miles south of east of Currie Well.

GOLD MOUNTAIN RIDGE AND THE HILLS TO THE NORTH OF GRAPEVINE CANYON.

TOPOGRAPHY AND GEOGRAPHY.

Gold Mountain, with an elevation of 8,145 feet, is situated on a ridge which parallels Slate Ridge. North of Tokop the two ridges are connected by low hills. The Gold Mountain ridge lies to the south of Oriental Wash and to the east of Death Valley. In the vicinity of Gold Mountain the ridge is rugged and steep, being formed of sharp peaks and V-shaped valleys. Strikingly distinct from the other summits is the basalt-capped butte northeast of Old Camp. A number of northeast-southwest ridges northwest of Grapevine Canyon are joined to Gold Mountain on the south or are isolated in the desert gravels near by. These ridges, which parallel the trend of the Gold Mountain ridge, are here included in its description. On the other hand, the mountain front north of Grapevine Canyon, northeast of Death Valley, and southwest of the California-Nevada line is structurally an extension of the Amargosa Range.

The distributive drainage characteristic of arid regions is well exemplified in these hills. Fragmental material from the Gold Mountain ridge near the Rattlesnake mine has reached lower levels by three distinct channels. A portion of the gravels moved to the southeast and found lodgment either around the playa in Grapevine Canyon or was washed out into Sarcobatus Flat near Thorp's mill. A second portion traveled south 4 miles, north of east 2 miles, and then south around the hill at the junction of the county line and parallel 37° 15′. The material eventually mingled with that which took the more direct route already described. A third portion did not swing around the hill just mentioned, but continued north of east into Sarcobatus Flat. Probably at any one time but a single channel was used, but in comparatively recent times all three have been used.

---

[a] Ransome, F. L., Bull. U. S. Geol. Survey No. 303 (in press).

Juniper and piñon cover the crest of the Gold Mountain ridge, the timber line being on the 6,200-foot contour on the south and on the 6,700-foot contour on the north side of the ridge. Yucca groves are sporadically distributed over the rhyolite and basalt flows to the south of the main ridge, while a fair growth of grass covers Gold Mountain and the surrounding hills. The main ridge on which Gold Mountain is situated is well watered. Tunnels and wells in the vicinity of Old Camp strike water at depths varying from 50 to 75 feet. The water in some cases appears to flow from crevices in granite and in others to seep from the overlying soil. Two wells located in Cambrian sedimentary rocks 1 mile east and northeast of Tokop are about 20 feet deep. The water is said to come from rock crevices. The water of Sand Spring, west of the boundary of the area here discussed, in Death Valley, is rather strongly impregnated with sulphur, but is nevertheless usable.

GENERAL GEOLOGY.

The formations exposed in the Gold Mountain ridge and the adjacent hills, from the base up, are as follows: Cambrian sedimentary rocks, quartz-monzonite porphyry, post-Jurassic granite, pre-Tertiary diorite porphyry, rhyolite, and basalt.

SEDIMENTARY ROCKS.

Cambrian.—Cambrian rocks form the rugged ridge northeast of Tokop and the western front of the ridge facing Death Valley. A thin band of Cambrian rocks rims the southern border of the granite batholith which forms the central portion of the main ridge. These narrow Cambrian masses between the granite and the Tertiary rhyolite are due to the fact that resistance to late Mesozoic and early Tertiary erosion was greater in the granite and the adjacent metamorphosed Cambrian rocks than in the unmetamorphosed Cambrian rocks. Large inclusions of Cambrian schist and metamorphosed limestone are sporadically distributed in the granite, while small fragments are widely distributed in both granite and rhyolite.

The Cambrian rocks of the Gold Mountain ridge include interbedded metamorphosed shales and impure limestones and calcareous sediments. The amount of metamorphism suffered by these rocks decreases with increase of distance from the granite batholith. Argillaceous rocks are widely distributed, although they do not appear to be present at the extreme west end of the range. They include fine-grained, well-banded slates of black to light-gray color and biotite schists both with and without knotlike aggregates of biotite.

Limestone and calcareous sediments are represented by white and gray marbles of medium to coarse grain and by lime-silicate rocks in which brown garnet, vesuvianite, epidote, tremolite, chlorite, and serpentine are developed. These rocks are usually well banded. One of the striking types is composed of layers of brown garnet and white quartz, alternating with fine-grained bands of quartz, epidote, and other silicate minerals in which large crystals of brown garnet are locally embedded. Garnet also occurs in veins cutting the rocks, indicating a second recrystallization of that mineral. A single specimen of the banded rock examined under the microscope proved to be an uneven-grained, banded rock. The contemporaneous constituents of recrystallization include slightly brownish garnets, quartz, epidote, zoisite, tremolite, calcite, orthoclase, and plagioclase (oligoclase and andesine), in places showing Carlsbad twinning. Apatite and titanite are present as accessories and sericite and chlorite as secondary minerals derived from the feldspars. At the west end of the Gold Mountain ridge bands and lenses of brown garnet rock up to 30 feet in thickness are interbedded with white and canary-yellow marble. Crystals of garnet also occur in the numerous interstices of the weathered surface.

These metamorphosed sedimentary rocks are similar to those of Slate Ridge, which grade into less metamorphosed facies, probably of Lower Cambrian age.

IGNEOUS ROCKS.

Quartz-monzonite porphyry.—The granite, particularly near Old Camp, contains inclusions up to 1 foot in diameter of a fine-grained, medium-gray igneous rock, formed of biotite and white feldspar, in which are rather sparse phenocrysts of pink orthoclase reaching a maximum length of one-half inch, smoky quartz grains one-fourth inch in diameter, and smaller biotite plates. Under the microscope the groundmass is seen to be hypidiomorphic granular and to consist of grains of orthoclase and partial crystals of plagioclase (oligoclase) and biotite, with a little quartz. Magnetite, titanite, zircon, and apatite are abundant accessory minerals. The phenocrysts are orthoclase, much of it zonally grown, quartz, biotite, and more rarely plagioclase. Similar inclusions occur in a number of the post-Jurassic granular rocks in other portions of the area, but beyond the fact that the quartz-monzonite porphyry is older than the granite, its age is unknown.

Post-Jurassic granite.—The central portion of the Gold Mountain ridge is a mass of granite which prior to the formation of Oriental Wash was probably connected with the granite batholith at the west end of Slate Ridge. The smaller area of granite 8 miles west of Gold Mountain was also perhaps at one time connected with the Slate Ridge batholith. Dikes and apophyses extend from the main mass into the Cambrian rocks.

The predominant type in the vicinity of Gold Mountain is an uneven-grained, coarse, pink biotite granite. The constituents are pink feldspar, locally in good crystals one-half inch long which show Carlsbad twinning, slightly smoky quartz, and black biotite. Microscopic examination shows this granite to be formed of quartz, microcline, microperthitic orthoclase (in some specimens), oligoclase, biotite, and magnetite. Quartz shows strong undulose extinction. Sericite and kaolin are alteration products of the feldspars. The granite at the west end of the ridge is more variable in character, the most abundant type being a fine- to medium-grained gray biotite granite. This is composed of quartz, orthoclase, plagioclase, and biotite, with titanite, magnetite, muscovite, zircon, and apatite as accessory minerals. Plagioclase (oligoclase) occurs in rude laths. The coarse pink variety of Gold Mountain is also present, and it grades into a gray granite porphyry, with fine-grained groundmass of quartz, biotite, and feldspar. The phenocrysts, which reach a maximum length of one-half inch, are tabular gray feldspars rimmed by white bands, some of them Carlsbad twinned; rounded, strongly smoky quartz, and hexagonal biotite. The microscope shows that this granite porphyry has a hypidiomorphic groundmass, composed of orthoclase, quartz, biotite, and plagioclase, with magnetite, apatite, and titanite as accessory minerals. Quartz in this granite is notably variable, being abundant in some outcrops and inconspicuous in others. With decrease in quartz the normal granite porphyry grades into a quartz-poor hornblende-bearing porphyry in which some pink feldspar phenocrysts are 1 inch long. Under the microscope this proves to be a quartz-monzonite porphyry, with rare tabular microcline and orthoclase and lath-shaped plagioclase phenocrysts. The groundmass, which is allotriomorphic granular, is composed of quartz, biotite, hornblende, orthoclase, and plagioclase. The accessory minerals are titanite, apatite, zircon, and magnetite. A little epidote is secondary to hornblende and biotite.

Both areas of granite are cut by dikes of a fine-grained pink apatite, in some places as rich in biotite as the granite and in others poor in this mineral. The granite is also cut by coarse pegmatite dikes and grades into similar pegmatitic masses. Graphic granite is not uncommon and at many points is the transition facies between granite and coarsely crystalline pegmatite. The aplite is, as a rule, older than the pegmatite, although in one case an aplitic dike passes along its strike into a dike with narrow aplitic border and pegmatitic center.

The yellowish outcrops of granite at the base of the ridge are discontinuous, being, as a rule, mere heaps of bowlders partially buried in granite soil. The outcrops near the crest are much more continuous. The granite intrudes Cambrian sediments, and rhyolite flows

of Miocene age lie upon its eroded surface.   Its age is in consequence
post-Cambrian and pre-Tertiary.   The granite is also cut by dikes of
diorite porphyry presumably of pre-Tertiary age.   It is probably of
post-Jurassic age.

Diorite porphyry.—Dikes of diorite porphyry reaching a maximum
observed width of 40 feet cut the Cambrian rocks and the granite of
the Gold Mountain ridge.   The rock is greenish in color, contains
abundant phenocrysts of white altered feldspar, and tends to weather
into spheroidal bowlders.   It is probably of pre-Tertiary age.

Rhyolite.—The hills to the southwest of Gold Mountain, the ridge
east of the Rattlesnake mine, a number of isolated hills and ridges to
the east, south, and southwest of the latter ridge, and a small area
northwest of the granite at the west end of the Gold Mountain ridge
are composed of rhyolite.   The presence of a small outlier of rhyolite
on granite, three-fourths of a mile southeast of Tokop, indicates that
erosion has removed considerable portions of a once more extensive
rhyolite flow.

The most widely distributed type is a lilac-gray rock of lithoidal,
in places rather incoherent, groundmass.   The medium-sized phenocrysts, which in bulk usually equal the groundmass, are tabular crystals of glassy, unstriated feldspars, often showing blue and purple
color plays, and rounded grains of colorless quartz.   Smaller plates
of biotite are locally present.   Pink, purple, and red facies also occur.
Black glassy types in which feldspar greatly predominates over
quartz, to judge from the hand specimen, may be more properly called
latite.

Many of the rhyolite flows are flow breccias, containing abundant
inclusions of rhyolite of different colors and textures.   Vesicular
facies showing flow lines are unusual.   Interbedded with the rhyolites
are sandy beds which contain normal feldspar and quartz crystals
and rounded rhyolite pebbles.   These are contemporaneous tuffs of
either aqueous or subaerial origin.   It is probable that minor flows
of basalt were contemporaneous with those of rhyolite, since vesicular
basalt fragments are in places included in rhyolite.

The rhyolite occurs in mesa-like ridges whose top is often coincident with the surface of a resistant rhyolite flow.   This rock lies
unconformably upon granite and Cambrian limestone, and upon its
slightly eroded surface basalt flows rest in turn.   The rhyolite is to
be correlated with that of the Amargosa and Kawich ranges and the
Bullfrog Hills and is probably of early Miocene age.

Basalt.—Basalt covers the butte between Old Camp and Tokop and
occurs in two small areas to the west of the butte and in a number of
small areas 11 miles south of Gold Mountain, resting upon the slightly
eroded rhyolite surface.   The basalt of the butte just mentioned is
a dense dark-gray flow rock, which is vesicular along some bands.

This rock, like some of the basalt of the Panamint Range, is spotted by areas of lighter color. The phenocrysts, which are small and fairly abundant, include striated feldspar, glassy green olivine, and blackish-green pyroxene. The basalt breaks down into rounded bowlders heavily stained by iron oxide. It is later than the rhyolite and is probably contemporaneous with the basalt of Slate Ridge and Mount Jackson and of late Pliocene or early Pleistocene age.

STRUCTURE.

The Cambrian rocks dip steeply away from the granite batholith, and near-by faults and brecciation, isoclinal folds, and buckled strata are developed. (See fig. 16.) At a distance from the granite the dips become less pronounced, but the mass of granite is so large and the Cambrian masses so small that rather high dips are characteristic throughout them. The granite is cut by fault and joint planes. Prior to the outflow of the rhyolite and basalt the granite and Cambrian rocks had a topography somewhat less rugged than at present, but the Gold Mountain ridge appears to have existed. Blocks of rhyolite north of Grapevine Canyon are tilted in different directions by normal faults, the majority of which strike northeast and southwest.

ECONOMIC GEOLOGY.

The Gold Mountain mining district, organized January 25, 1868, embraces within its limits the Gold Mountain and Slate ridges. Mines have been worked in the district intermittently since that time, and several mills have been built, none of which are now in operation. Old residents estimate the total product of the distrist as $500,000, the concentrates being hauled by wagons to Belmont and Austin, Nev. The mines and prospects lie in the more highly metamorphosed Cambrian rocks and in granite, and these rocks in this vicinity are probably most favorable for prospecting. The contact of the rhyolite and granite 9 miles south of west of Gold Mountain is in places altered to an iron-stained, incoherent mass, cemented by chalcedonic quartz, and this contact is perhaps worthy of examination.

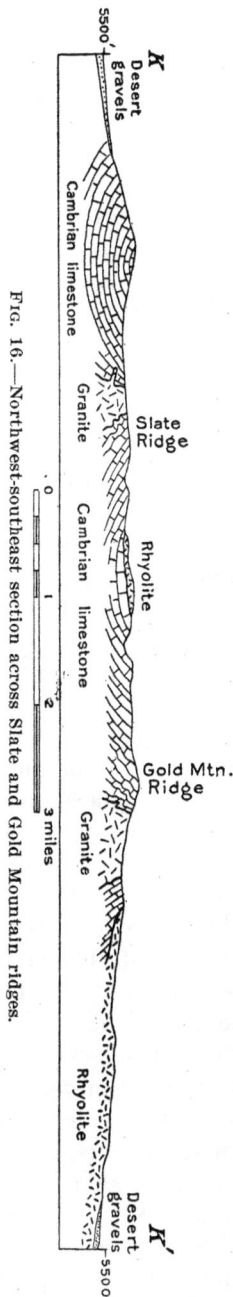

FIG. 16.—Northwest-southeast section across Slate and Gold Mountain ridges.

The almost deserted village of Old Camp, near Gold Mountain, is situated 30 miles west of south of Goldfield and 20 miles southeast of Lida. Abandoned mines and prospects are numerous in the vicinity and a number of prospects are being developed 2 or 3 miles south of the village.

The Central mine, which supplied the ore used in the mill at Old Camp, may be taken as a type of ore deposits in the granite of this region. This mine is situated on the side of a deep gulch 1½ miles north of east of the mill. Five tunnels, with an average length of 300 feet, pierce the granite and all are situated on a single vein or system of connecting veins. The feldspars of the granite within 20 feet of the vein are considerably kaolinized and the biotite is bleached or altered to a sericitic mineral. The quartz vein, or rather zone, within which the numerous connecting quartz veinlets and stringers occur, is from 1½ to 6 feet wide, the proportional amount of quartz increasing with the narrowing of the zone. The ordinary veinlets are from 2 to 5 inches wide. The zone also contains many ellipsoidal and globular areas of quartz which, at least in the plane of observation, are independent masses. The quartz zone in some places changes its direction 90° within 100 feet. While minor postmineral faults are common, the displacement ranging from 6 inches to 7 feet, the major changes in strike and dip are evidently parts of the original structure.

The quartz is white and translucent or slightly smoky, with a strong vitreous luster. Vugs with inch-long crystals are rather rare. In places the clear quartz seems to grade into a gray chalcedonic form, the deposition of which by ordinary waters can scarcely be doubted. Isolated crystals of pyrite, with very rare crystals of chalcopyrite and galena, are embedded in the quartz, while the former abundance of pyrite in particular is shown by numerous iron-stained cavities of cubical form. Both hematite and limonite occur in the porous quartz, and where these are abundant gold values rise. An occasional malachite stain and a cerussite coating are also present. Dendrites of manganese dioxide occur in the granite, and are probably derived from the alteration of the granite rather than from a decomposing gangue mineral. Horn silver is reported to be present in small amounts, but was not seen. The ore is free-milling, the arrastres, now abandoned, having saved about 75 per cent of the assay value. Films of a bluish-white chalcedony were noted at a number of places. Some of it was deposited prior to the oxidation of the pyrite and some was deposited after partial or complete oxidation.

While the quartz is rather similar to the quartz of pegmatitic origin in the district, and the form of the ore deposit rather suggests that of a pegmatitic dike, the apparent gradation into chalcedony indicates a less close connection with waters of the granitic magma. Water appears to have filled a most complex zone of fracture with quartz and sulphides, the gold probably being originally contained in pyrite. Later oxidizing waters attacked the sulphides and altered them to oxides and carbonates, setting the gold free. These waters apparently carried some silica in solution, which was deposited as chalcedony.

The mines in the vicinity of Old Camp are situated near water sufficient for mining and milling purposes. Timber for fuel and mining use is standing within 2 or 3 miles. The railroad terminus at Goldfield is 35 miles distant.

TOKOP.

Tokop is 4 miles in an air line northeast of Old Camp and 25 miles by road west of south of Goldfield. The properties of the Gold Crest Mining Company, situated three-fourths of a mile south of Tokop, were examined in some detail. On the Ouida claim a vertical vein cuts the Cambrian garnet-quartz-epidote rock, which strikes N. 10° E. and dips 20° W. This vein is 1½ feet wide and strikes N. 80° W. The quartz of the vein, which is white and semitransparent, has been intensely crushed, and the fragments have been cemented by hematite and limonite. Dendrites of manganese oxide also occur. Some portions of the quartz are compact, with here and there an iron-pyrite cube unaltered, while other portions have a few vugs lined with quartz crystals or are honeycombed with iron-pyrite casts. From the fact that the compact varieties are refractory and the honeycombed varieties are free-milling, it is evident that the gold was set free by the alteration of the pyrite. A mashed greasy phase of the garnet-quartz-epidote rock on one side of the vein is said to pan, but whether pyrite originally occurred in the metamorphosed limestone or whether the values were derived from the surface alteration of the quartz vein is unknown.

Angular fragments of altered limestone and shale are common near the borders of some veins. In such places silicification has extended into the limestone, and these belts, like the quartz veins, weather in relief. Such veins grade into sheared and brecciated zones of silicified limestone, and these again are reported to carry values.

Another vein is similar to the vertical vein first described, but the shattered quartz is cemented by a gray chalcedonic quartz. Still another, at first sight, appears crustified, but close inspection shows that the appearance is due to a longitudinal fracturing of the vein and subsequent filling by limonite and chalcedonic quartz. Both forms of quartz are said to assay, but the earlier quartz appears to carry both the sulphides already mentioned and slight amounts of galena as noted in other veins. It may be that the chalcedonic quartz appears to carry values simply because sufficient care has not been taken to eliminate the older quartz from samples.

Limonite, hematite, cerussite, free gold, and a little malachite were observed as secondary minerals. Probably some chalcopyrite exists at greater depths. One of the last changes which these ore deposits, like those of Old Camp, have undergone is the local deposition of a thin film of bluish-white chalcedony.

The contact of the granite and limestone here probably has locally a rather gentle dip, and deep mines in limestone near the contact may encounter granite. Mr. Joseph Mackedon, the manager of the Gold Crest Mining Company, states that in several prospect holes the quartz veins have past from metamorphosed limestone to granite without diminution in size or value. From the striking resemblance of these veins to those of Old Camp this is to be expected.

Similar veins of iron-stained quartz carrying gold values occur at a number of places in metamorphosed Cambrian rocks around Tokop.

Waters carrying silica and metallic salts in solution appear to have deposited these substances in strong fractures which extend, in some cases, at least, into the granite. The country rock was to a less extent impregnated. The quartz veins were subsequently crushed and iron oxide and chalcedonic silica were deposited in the fractures. Simultaneously, probably, iron pyrite was dissolved and the gold set free.

Water and wood for mining and domestic purposes are near at hand, and the distance to the railroad at Goldfield is 25 miles. The Rattlesnake mine, near the properties described, appears to be in the same rock formation, and its ore deposits and those of Tokop are said to be similar. Eight or ten years ago the Rattlesnake produced $150,000, and recently it has been reopened. The veins of Tokop, like those of Old Camp, seem strong and will probably be permanent to such depths as mining is possible. With depth, however, the ore will become refractory and may become leaner.

ORIENTAL WASH.

Several prospects, now abandoned, are located in the altered limestone west of the granite area at the west end of the Gold Mountain ridge. At one prospect a thin quartz vein cuts the limestone. Malachite, azurite, chrysocolla, and limonite-stained chalcedonic quartz occur in irregular patches and veinlets through the quartz. Vugs occur in the secondary minerals, and the azurite and malachite are in places covered with numerous quartz crystals. A little chalcopyrite is embedded in the oldest quartz and cores of chalcopyrite are surrounded by the secondary copper minerals. A vein of malachite 1 inch in width, apparently replacing the limestone, is exposed in another prospect. At still another prospect, in association with malachite, azurite, and coral-red opaline quartz, heavily stained by manganese dioxide, is a streak from 1½ to 2 inches wide of a canary-yellow granular mineral which appears to replace limestone. Dr. W. F. Hillebrand determined this mineral to be chloropal, a hydrous iron silicate. The general resemblance in structural relations and mineralogic composition of these deposits to those near the granite of the Belted Range at Oak Spring is worthy of note. The nearest water to these prospects is at Sand Spring, 6 miles away, and fuel can be obtained from Gold Mountain within 10 miles. By road the prospects are 45 miles from the railroad terminus at Goldfield.

SLATE RIDGE.

TOPOGRAPHY AND GEOGRAPHY.

Slate Ridge lies between the valley south of Mount Jackson on the north and Oriental Wash and an opposed westward-reaching arm of Sarcobatus Flat on the south. The ridge trends a little north of east, and near its middle is almost severed by opposed gravel-filled valleys. West of the Stateline Mill the ridge is formed of a number of low domelike granite hills, which reach a maximum elevation of 6,500 feet. The hills are rather gently sloping and without prominent rock exposures. From a distance they are yellowish gray in color. East of the Stateline Mill the rocks are largely of sedimentary origin and there is a sharp crest line. The east end of Slate Ridge is a mesa of lava flows which have been considerably dissected. The range is covered by a sparse growth of tree yucca and a little grass grows throughout its extent.

GENERAL GEOLOGY.

The formations exposed in Slate Ridge, from the oldest to the youngest, are the following: Cambrian sedimentary rocks, post-Jurassic granite, pre-Tertiary quartz-monzonite porphyry, pre-Tertiary diorite porphyry, earlier rhyolite, later rhyolite, and basalt.

SEDIMENTARY ROCKS.

Cambrian.—Cambrian rocks form the central part of Slate Ridge and occur in numerous areas isolated in the granite in its west end. Of the latter some are clearly inclusions buoyed up by the granite magma, while others may perhaps be connected with larger masses beneath the surface. Small Cambrian fragments are abundantly included in granite and the earlier rhyolite, particularly near the contact of these rocks with the sediments.

The Cambrian rocks of Slate Ridge are little metamorphosed on the road between Tokop and Goldfield. Here the interbedded series consists of paper-thin olive-green shales, gray limestones, or mag-

nesian limestones, usually rather thin bedded, and some impure medium-grained quartzites of purple, gray, or white color. Rounded
hills of gentle slope are developed in these slightly metamorphosed
rocks. The Cambrian rocks are intensely metamorphosed near
granite, and these facies have been described in the section on the
Gold Mountain ridge (p. 183), along which they are more extensively
developed.

No fossils were found in these rocks, but from their lithologic similarity to the rocks in the vicinity of Lida and Cuprite they are without much doubt of Lower Cambrian age.

Later tuffs.—Beneath the later rhyolite at several places is exposed
a thin band of white tuffaceous sandstones, possibly to be correlated
with the later tuffs of the Goldfield hills and Pahute Mesa.

IGNEOUS ROCKS.

Post-Jurassic granite.—Granite forms a considerable batholith on
the west end of Slate Ridge and smaller intrusive masses and dikes
occur in the Cambrian rocks. This rock is a coarse-grained biotite
granite of pinkish tone. The pink or white feldspar and slightly
smoky quartz individuals reach a maximum diameter of one-half
inch, while the black biotite plates are smaller. The granite tends to
weather into spherical masses. The surface of these bowlders appears to undergo a kind of cementation, and on further weathering
the hardened surface protrudes beyond the soft interior, forming
mushroom-like forms. The granite breaks down into a soil of the
constituent minerals and mechanical disintegration is so rapid that
recent detrital deposits extend as broad basins well up into the hills.
The granite is cut by parallel sheeting in certain localities and on
weathering assumes the appearance of a sedimentary rock. Microscopic examination shows the texture to be allotriomorphic. The
feldspar is orthoclase, considerably kaolinized and sericitized. Zircon and magnetite are present as accessory minerals.

Dikes of fine-grained biotite aplite, rather more siliceous than the
granite, cut it, and these on weathering protrude from the granite
mass. The granite is also cut by dikes and in turn grades into irregular masses of coarse-grained pegmatite. Some quartz veins are evidently of pegmatitic origin, since they grade into less siliceous pegmatites. In some instances microscopic examination shows quartz
individuals at the contact of granite and pegmatite to be common to
both rocks. At the Bullfrog-George mine fluorite and molybdenite
occur in a quartz vein which is without much doubt of pegmatitic
origin. Molybdenite occurs sporadically in small tablets and irregular areas in and between the quartz individuals. A bright-yellow
mineral in minute crystals and tufted aggregates, apparently secondary to molybdenite, was determined by Dr. Waldemar T. Schaller

to be molybdite (molybdenum trioxide). Purple fluorite occurs in crevices in the quartz, and fluorite cubes one-fourth inch in diameter also line vugs in the quartz. Microscopic examination shows that the quartz individuals in contact with fluorite possess sharp outlines, showing that they were but little corroded by the introduction of the fluorite, which is probably of pneumatolitic origin.

The granite is similar lithologically to that of Gold Mountain and has like relations to the Cambrian rocks and the Tertiary rhyolite. It is probably of post-Jurassic age.

Pre-Tertiary quartz-monzonite porphyry.—The Cambrian sedimentary rocks on the road from Tokop to Goldfield are cut by dikes which reach a maximum width of 40 feet. The rock is white and dense and contains a few feldspar phenocrysts. It is similar to the quartz-monzonite porphyry already described from the vicinity of Lida and, like it, is probably genetically related to the granite.

Pre-Tertiary diorite porphyry.—The Cambrian rocks and the granite are cut by narrow dikes of a greenish-gray much-altered rock in which some of the prominent kaolinized feldspar phenocrysts are one-half inch long. This is probably the pre-Tertiary diorite porphyry.

Earlier rhyolite.—The earlier rhyolite forms a considerable portion of the east end of Slate Ridge, and small areas of it occur, widely distributed over the central and western portions. In some places east of the Bullfrog-Goldfield road it appears from a distance to protrude from the later rhyolite. The earlier rhyolite is a flow rock of purple, gray, or white color. The groundmass, which is usually lithoidal, is equaled in bulk by the medium-sized phenocrysts of glassy unstriated feldspar, quartz, and biotite. Semituffaceous layers are interbedded with the flows northeast of Tokop. The earlier rhyolite lies upon the eroded surface of the Cambrian rocks and the granites, and in turn is overlain by basalt and the younger rhyolite. It is similar to that of the Amargosa Range and is probably of earlier Miocene age.

Later rhyolite.—The dissected mesa at the east end of Slate Ridge and a number of outliers in the recent gravels east of it are formed of flows of rhyolitic rocks. This later rhyolite has a dense groundmass of deep-brown color, in which are abundant glassy feldspar phenocrysts; in some facies very small and in others medium sized. Microscopic examination proves the groundmass to be a brown glass with well-developed flow lines. The phenocrysts are orthoclase, somewhat corroded, rare and small greenish augite crystals, and very rare quartz grains. While this rock is tentatively here called a rhyolite, chemical analysis might prove it to be of less acidic composition.

Beneath this rhyolite are tuffaceous sandstones in which are rhyolite pebbles. This is probably the later tuff (Pliocene) of the Goldfield hills. The rhyolite immediately above the sandstone is in many places vesicular and is a flow breccia.

This rhyolite is similar lithologically to the later rhyolite of the Goldfield hills, and the two have suffered equal deformation. They are without much doubt contemporaneous lava flows of Pliocene age.

Basalt.—Black vesicular basalt occurs in a number of small outliers upon the Cambrian rocks, the granite, and the earlier rhyolite and as small buttes on the border of Slate Ridge. The basalt is similar lithologically to that of Mount Jackson and, like it, is probably of Pliocene or early Pleistocene age.

STRUCTURE.

The Cambrian rocks at a distance from granitic intrusions lie in gentle folds many of whose axes trend northeast. The folding is comparable in intensity with that at Cuprite and Lida. On the other hand, they dip steeply away from the granite batholith and in its vicinity are buckled and minor isoclinal folds are common. (See fig. 16, p. 187.) It is evident that in this region the Cambrian rocks were folded prior to the intrusion of the granite, a process which superimposed upon the gentle folds complex elements. The granite is cut by normal faults. Since the Tertiary rocks were formed the ridge has suffered domical uplift, centering 4 or 5 miles east of the Stateline Mill.

ECONOMIC GEOLOGY.

Quartz veins, some of which are of pegmatitic origin, cut granite and the more metamorphosed Cambrian rocks. The best prospecting ground is the granite and the metamorphosed limestone and shales in its vicinity. In the late sixties mines were opened on Slate Ridge and considerable ore bodies were removed. The ledge at the Stateline Mill, now abandoned, is said to have been 20 feet wide, the ore averaging $40 per ton in gold. Recently a number of claims have been located in Slate Ridge and development work is now being done.

The Bullfrog-George prospect is situated on the side of a domical granite hill, near the Lida-Old Camp road. The surrounding hills are cut by quartz veins which weather in relief and can be traced for long distances. Some of these contain feldspar and others grade into pegmatite dikes. The Bullfrog-George claims are situated on a quartz vein from 4 to 9 feet wide, which is traceable for about a quarter of a mile. The vein is vertical and strikes N. 70° W. The contact with the granite is in some places gradational, the white translucent quartz of the granite passing into that of the vein without break. In other places the granite appears to have been shattered prior to the deposition of the quartz, which now fills linked cavities in the granite. The feldspars of the granite within 4 feet of the vein are locally much kaolinized. Apparently isolated in the quartz are small areas of pyrite and chalcopyrite, with less galena and chalcocite or a related sulphide. The quartz in many portions is intensely crushed, this crushing perhaps being contemporaneous with faults which cut the diorite-porphyry dikes of the vicinity. The crushed fragments have been recemented by limonite or a chalcedonic quartz intensely stained by limonite. With these knife-edges of limonite and in limonite-stained cavities malachite, cerussite, and traces of azurite occur. Such quartz pans free gold, and coarse pannings were examined from the heavily iron-stained contact of the granite and quartz. At the east end of the quartz veins less shattering was noted. Vugs lined with quartz crystals are common, and here the fluorite and molybdenite already described (see p. 192) occur.

The quartz vein itself appears to be a pegmatite which crystallized while portions of the granite were still viscous and other portions were comparatively solid. Later, faulting occurred and the vein was crushed. Since then limonite and chalcedonic quartz have recemented the quartz fragments. The period at which the sulphide mineralization occurred is unknown. The molybdenite and fluorite are probably of pegmatitic origin, while the sulphides were doubtless deposited later. Similar quartz from other prospects on the ridge was examined. Where iron stained it is said to carry free gold.

Water is at present hauled from a pipe line on the Lida-Old Camp road, 9 miles distant. The nearest timber grows in the vicinity of Old Camp. The prospect is 30 miles from the railroad terminus at Goldfield.

DEATH VALLEY.

TOPOGRAPHY AND GEOGRAPHY.

Death Valley is a deep depression lying between the Panamint Range on the west and the Amargosa Range on the east. The valley, which has a length of 120 miles and a width varying from 3 to 10 miles, unites with the Amargosa Desert at the south end of the Funeral Mountains. In the fifties a band of eighty emigrants are said to have perished here and the valley received its name from this tragedy. In the Pahute language the valley is called Tomesha, meaning "ground afire." Many fantastic stories, most of them wholly without foundation, center in Death Valley. Lives have been lost from year to year, but the majority of such sacrifices have been due to ignorance of desert conditions. In July and August the temperature is reported, apparently on good authority, to reach 136° F., and in summer none but the most desert-hardened men should enter the valley. In winter and fall the climate is delightful. In those portions of the valley which lie beneath sea level the dry air seems heavy and the brightest days are sultry. In the lower portions of the valley no animals other than a few coyotes live, and the oppressive silence is unbroken. Much of the valley is absolutely without vegetation and the adjacent mountain slopes are equally bare. Mesquite grows on the sand area west of Surveyors and Stovepipe wells and at Mesquite Spring. The upper portions of the alluvial slopes have here and there a sparse growth of creosote bush and white sage, and salt grass covers restricted areas around the wells.

The lowest point of Death Valley within the area mapped is about 280 feet below the sea level, but 15 miles farther south the depression is at least 125 feet deeper. The mountain ranges on either side of the valley are from 6,000 to 8,900 feet high, and in consequence of this great difference in elevation the slopes are steep and wide alluvial cones sweep up to the mouths of the deeply scored canyons. The floor of the valley is diversified by numerous hills and low ridges of Tertiary sediments. The part of the valley lying north of a low ridge at the head of Salt Creek is sometimes called Mesquite Flat.

A large sand dune occupies the bottom of Mesquite Flat. Most of the larger sand dunes are without vegetation, while some of the smaller ones (20 feet high and 100 feet in diameter) owe their existence chiefly to the protecting influence of mesquite and similar vegetation (Pl. II, B). Death Valley is the only portion of the area surveyed in which sand storms endanger life. The area mapped as sand contains within its limits many small areas of clay and is without doubt partially under water in times of excessive floods. In like manner the playa deposits mapped to the south, north, and east of the sand are studded with minor sand dunes.

South of Salt Creek the center of the valley is a loamy flat covered with salt grass. This passes gradually southward into a salt marsh, 4 miles wide, which, according to Campbell,[a] extends about 25 miles south of the area mapped. It is a dirt-brown flat, containing numerous channels and ponds of stagnant or gently flowing salt water. The soil itself is heavily impregnated with salts, and by fractional recrystallization the margins of the water bodies are coated with glittering white salt deposits. The salt-cemented soil stands up in rugged pinnacles and hummocks from 6 to 18 inches high. Campbell[b] states that a specimen of this material from a point south of Furnace Creek has the following composition:

[a] Campbell, M. R., Bull. U. S. Geol. Survey No. 200, 1902, p. 18.
[b] Loc. cit.

Analysis of soil from Death Valley.

	Per cent.
Chloride of sodium	94.54
Chloride of potassium	.31
Sulphate of sodium	3.53
Sulphate of calcium (hydrous)	.79
Moisture	.14
Undissolved residue (gypsum and clay)	.50
	99.81

This large supply of salt, while at present, as Campbell states, too impure to be of commercial value, may in the future be economically important. From 1883 to 1887 an extensive plant extracted borax from the salts of this marsh, but on the discovery of extensive colemanite deposits in the Tertiary lake beds on Furnace Creek it was closed down. Borax is also reported in the clay of the plava 2 miles east of south of Surveyors Well.

Death Valley is one of the best watered areas within the limits of the area here discussed, and the water for the most part is good. Salt Creek and the warm Indian Springs have already been described (pp. 19, 20). At Stovepipe Wells, Surveyors Well, Ruiz Well, Salt Creek Wells, and the water hole on the Furnace Creek road, 1 mile north of the boundary of this area, water stands in shallow holes dug in the sand and clay. The water of the last-named hole is too salt for man's use, although animals will drink it. Mesquite Spring is a small spring in Recent gravels, while the water of Cow Creek and Triangle Spring flows from Tertiary lake beds.

GENERAL GEOLOGY.

SEDIMENTS.

General statement.—Two sedimentary series much later in age than the Paleozoic rocks of the Amargosa and Panamint ranges outcrop in Death Valley. The one is composed of coarser fragmental material, well rounded, and appears to have been deposited in a wave-swept lake, while the material of the other is angular or subangular and evidently analogous in origin to the present alluvial slopes and fans. Associated with each are finer grained clays and limestones— those of the one series, lake deposits; those of the other, playa deposits. The delineation of these finer deposits on the map is in many cases probably inaccurate. The lake deposits are clearly older than those of the playa and are better cemented and in general more deformed. The two formations, while less extensively developed, are much more easily differentiated on the east side of the Amargosa Range and in the adjoining portions of the Amargosa Desert, where the detrital and playa deposits are uncemented angular gravels, clays, and chemically precipitated limestones, which are particularly undeformed, while the older lake beds are sandstones and well-rounded conglomerates which dip at high angles. Below Ash Meadows, in the Amargosa Desert, Campbell[a] found evidence that the detrital deposits were laid down after the folding and elevation of the lake beds into the Funeral Mountains. The time gap between the two is considerable, and the older beds are tentatively considered by Campbell[b] Eocene and the others Pliocene. It is believed, however, from evidence obtained in the Amargosa Range, that the older beds are probably largely as late as the Miocene and that they may be the shore deposits of Siebert Lake, while the presence of basalt, contemporaneous with the upper portion of the younger beds, indicates that their deposition probably extended into the Pleistocene. The later beds are evidently the equivalent of the older alluvium.

Territory lake beds (including the Siebert lake beds).—The Tertiary lake beds, which are a northwesterly extension of the folded Tertiary sediments extending across the Funeral Mountains southwest of Furnace Creek ranch, are confined, in the area surveyed, to the east side of Death Valley near the boundary. Numerous hills of these beds rise above the Recent gravels, and the scale of the map permits only an approximate representation of their complex distribution.

The Tertiary lake beds consist of white, yellow, and green consolidated clays, friable sandstones with ironstone concretions, rounded and subangular gravels, and thin limestone lenses. Much of the clay shows sun cracks and ripple marks, indicating that the lake was at times shallow and even dry. The subangular form of certain of the gravels indicates that cloudbursts at times spread sheets of detrital deposits over the lake beds. Colemanite and other chemical precipitates interbedded with the other deposits were laid down during periods of unusual evaporation. The more northerly hills to the west of the Daylight Spring-Furnace Creek road were

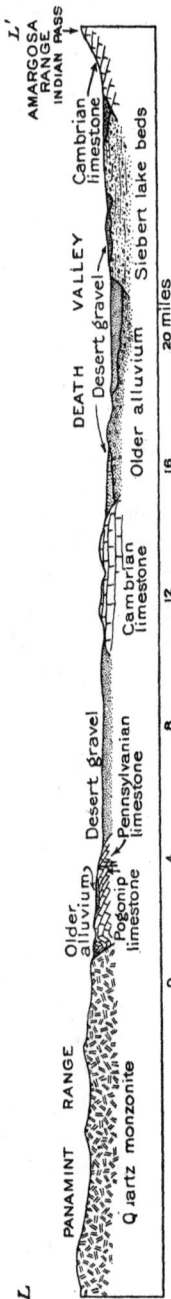

Fig. 17.—East-west section across Death Valley, 6 miles north of boundary of area mapped.

[a] Campbell, M. R., Bull. U. S. Geol. Survey No. 200, 1902, p. 16.
[b] Loc. cit.

not visited by the writer. but specimens collected by Mr. R. H. Chapman are reddish-brown conglomerates and conglomeratic sandstones with rounded pebbles. The pebbles, which reach a maximum diameter of 2½ inches, are limestone, quartzite, and schist derived from the Amargosa Range. They are probably members of the lake-bed series.

The Tertiary lake beds are folded into open folds, usually with northwest axes. (See fig. 17.) Joints from 4 to 5 feet apart are developed.

Older alluvium.—Hills of the older alluvium diversify the surface of Death Valley, at two places extend a considerable distance up the Amargosa Range, and form the north end of the Panamint Range. In the large sand area north of Salt Creek a few hillocks of slightly consolidated clay occur, and these probably indicate the former extension of these deposits across the valley.

The road from Daylight Spring to Stovepipe Wells passes through a desiccated ridge from 200 to 400 feet high, which trends northwest and southeast and is cut by antecedent drainage lines. Next to the Amargosa Range the hills are composed of slightly consolidated beds of angular or subangular bowlders and bands of gravel and sand. Cross-bedding and local unconformities are common. Well-developed joints form buttresses on the main cliffs, and large bowlders crown many of the high pinnacles. Within one-half mile of the western edge of the hills the sand beds are replaced along their strike by clay, and clay layers wedge in between the bowlder beds. At the western edge of the hills the base of the series is clay and the top bowlder beds. A section measured 2 miles south of east of Stovepipe Wells is as follows:

Section in Death Valley near Stovepipe Wells.

	Feet.
Angular bowlder beds, similar to those of the present alluvial slopes	150+
Clay and bowlder beds in equal development	110
Salmon-pink clay beds, from 3 inches to 3 feet thick ; a few layers of angular bowlders interbedded	100

At Triangle Spring fine-grained, dirty-brown cellular limestone is interbedded with the clays. The limestone shows desiccation cracks. Compact, fine-grained, light-colored limestone and caramel-brown mammillary gypsum also occur here. The beds exposed at Salt Creek include yellow, white, and green clays, with here and there a thin bed of white limestone. Brownish beds of angular bowlders constitute a considerable portion of the upper part of the section. The long ribbon of these deposits near the Grapevine Mountains northeast of Surveyors Well consists of tawny-yellow clays and thin beds of cellular limestone. At Grapevine Springs the deposits consist of limestones and fine clays. The clay contained crystal aggregates of selenite. The limestone is white or gray in color and is partly compact and very finely laminated and partly porous. The areas in the middle of the valley northwest of these springs are of interest, since flows of the supposed Pleistocene basalt are interbedded with the clay, while other layers contain bowlders of basalt. In some places the clay beneath the basalt has been reddened and slightly baked. These clay beds, which are well up in the section, are broadly contemporaneous with basalt eruption.

About 2,500 feet of the older alluvium is exposed in the Amargosa Range, and it is probable that the series was originally over 3,000 feet thick. The beds are gently flexed, and dips from the mountains to the valley axis predominate, particularly in the northern part of the valley. The sediments have been jointed, the principal system striking N. 25° W., and along some of these joints occur normal faults with from 2 to 4 feet displacement.

GEOLOGIC HISTORY.

In Tertiary, probably early and middle Miocene time, Death Valley did not exist, Amargosa and Panamint ranges were low, and their southern portions [a] at least were covered by a lake which extended well into the present Death Valley south of Salt Creek. This lake was at least partly contemporaneous and may have been connected with that in which the Siebert lake beds were deposited, and which probably at one time covered almost the entire area surveyed. In late Pliocene time, however, Death Valley was probably a closed basin, occupied by a sheet of water, which appears to have been a playa. This playa varied widely from time to time in extent and was bordered by the uplifted Panamint and Amargosa ranges. Before the deposition of the older alluvium ceased the inwash of material from the mountains raised the center of the valley hundreds of feet above the present level. Alluvial slopes, much like those of to-day, extended well up on the Panamint and Amargosa ranges. In the playa great thicknesses of clay were deposited, and at periods of unusual desiccation limestone and gypsum were precipitated. Unusual floods from time to time throughout the period of deposition spread thin layers of bowlders over the playa. Basalt outflows accompanied the deposition of the upper playa deposits.

Death Valley is believed to have been roughly blocked out in late Miocene and early Pliocene time, through erosion and deformation, including faulting and possibly folding. The folding of the Amargosa and Panamint ranges does not alone account for the valley, and it appears to be a block dropped down between the bounding ranges by faults. The earliest structural lines, probably of late Miocene age, trend northwest and southeast and include the fault [b] on the north-

---

[a] Spurr, J. E., Bull. U. S. Geol. Survey No. 208, 1903, pp. 189, 202.
[b] Campbell, M. R., Bull. U. S. Geol. Survey No. 200, 1902, p. 16.

east side of the lake beds in the Funeral Mountains and normal faults in the Panamint Range. In early Pleistocene time the lines of disturbance were north-northwest, and normal faults, with a strike parallel to the valley axis, occur in the older alluvium east of Stovepipe Wells. In the Panamint Range 22 miles beyond the boundary of the area mapped and in line with Death Valley a narrow block, apparently faulted into the older rocks, has, at a distance of 15 miles, the appearance of the older alluvium. The fronts of the Panamint and Amargosa ranges facing Death Valley are very steep and may be due to the erosional shifting of a fault scarp. The absence of Paleozoic inliers in the valley indicates the steep grade with which these rocks descend beneath the valley gravels. That the drainage lines of the Panamint and Amargosa ranges flowing into Death Valley have been revived in comparatively late geologic time is indicated by the intense dissection of the prebasaltic mature topography of the Panamint and Amargosa ranges. (See pp. 202 and 161.) The history of Death Valley has been complex, and the solution of the problem requires more detailed work than was possible in the present reconnaissance.

ECONOMIC GEOLOGY.

Extensive deposits of colemanite and other borax minerals occur in the Tertiary lake beds near Furnace Creek, and the extension of these beds in the area mapped is worthy of careful prospecting. " Marsh " borax has been found 2 miles south of Surveyors Well, and it is probable that this has been leached from the older alluvium near by. The clays and limestone of this formation may contain important colemanite deposits.

PANAMINT RANGE.

TOPOGRAPHY AND GEOGRAPHY.

The Panamint Range is 130 miles long and trends north-northwest parallel to the Sierra Nevada. At its north end lava mesas unite it with the Amargosa Range; its south end passes into low hills of Tertiary sediments and lavas capped by later basic volcanics.[a] The ascent from Death Valley on the east ranges from 6,000 to 11,000 feet, while that from Panamint and Termination valleys on the west is almost as great. The range culminates 20 miles south of the area here mapped in Telescope Peak, which the Wheeler survey determined to be 10,938 feet high. Within the area the highest point is Tin Mountain, whose elevation is 8,900 feet. The range crest is somewhat west of the center, and the western slope seems a succession of cliffs and declivities. The range is scored by deep canyons, some of which are comparable with the most famous gorges of the Rocky Mountains. Marble Canyon (Pl. III) is particularly impressive. The sharply

---

[a] Spurr, J. E., Bull. U. S. Geol. Survey No. 208, 1903, p. 201.

curving inner canyon, which is but 10 feet wide, is overhung by walls 200 feet high, and many portions of the stream channel are in perpetual shadow. The walls are beautifully smoothed and rounded by erosion 75 feet above the stream bed. Rock cliffs 10 to 100 feet high, which in other regions would form waterfalls, break the more gentle grades of gravel. The brilliant yellows, reds, whites, and blacks of the limestone enhance the beauty due to form.

The crest of the range is characterized by an older, more mature topography, with well-graded stream channels and gently rounded peaks and slopes. Near the southern boundary of the area mapped and for a distance of 4 miles south of Tin Mountain this belt is 3 miles wide, while the intermediate belt along the crest line is but 1 mile wide. Remnants occur in favorable positions on the middle slopes of the mountain range above an elevation of 5,000 feet. On the southern border of the area the basalt covers a portion of this surface—a fact which indicates that the surface was probably developed in Pliocene time. The older alluvium, in part, is formed of remnants of the alluvial slopes, the gravel pediment of this subdued mountain range.

Sparse growths of piñon and juniper cover the southwest corner of the area and the northeastern and southern slopes of Tin Mountain, while cottonwood and willow grow along Cottonwood Creek. Grass is found around the small playa northwest of Goldbelt Spring and in the valleys in the higher portion of the range. The Panamint Range, because of its elevation, has a heavier annual precipitation than most of the other ranges. One or more springs rise in nearly all the main gulches tributary to Cottonwood and Marble Canyon creeks. Several springs lie south of Tin Mountain. Emigrant Spring, to the south of the area mapped, flows 1,100 gallons of water per day.

GENERAL GEOLOGY.

The formations of the Panamint Range, from the oldest to the youngest, are as follows: Prospect Mountain quartzite (?), Pogonip limestone, Pennsylvanian limestone, earlier quartz-monzonite porphyry, post-Jurassic granitoid rocks (quartz monzonite, soda syenite, and granite), later quartz-monzonite porphyry, older alluvium, basalt, and Recent alluvium.

SEDIMENTARY ROCKS.

Prospect Mountain quartzite (?).—Tucki Mountain is formed of metamorphosed sediments identified with the supposed Prospect Mountain quartzite of the Amargosa Range. (See p. 162.)

Pogonip limestone.—The most widely distributed formation of the Panamint Range is the Pogonip limestone, of which at least 3,000 feet is exposed. The limestone, which is compact and fine grained, ranges in color from dark gray to black. The dark color is, at least in part, due to carbonaceous matter, since some beds are fetid, and when heated exude a tarry substance. The bedding is, as a rule, massive, although some slightly argillaceous facies are thin bedded. Cross-bedding was noted in several places. Near the top of the section occur layers and lenses of black, semitranslucent flint from one-half inch to 3 inches thick, separated by bands of limestone from 1 to 18 inches thick. In other beds globular and ellipsoidal flint concretions from one-fourth inch to 3 inches in diameter are abundant. A quartz-ite layer interbedded with the limestone occurs rather well up in the formation. The rugged exposures of this white or pinkish-white fine-grained quartzite outcrop on the western side of the playa 4 miles north of Goldbelt Spring. The bed of quartzite is 100 feet thick and has thin bands of limestone interbedded near its base, and especially near its top. Four miles south of Tin Mountain Mr. T. C. Spaulding noted interbedded in the limestone a 30-foot band of muscovite schist. The schist is less metamorphosed than that of Tucki Mountain.

At one locality the writer collected fragmentary fossils which Mr. E. O. Ulrich states include a shell belonging either to *Maclurea* or some related genus and a fragment of a pelecypod. These indicate undoubtedly Ordovician and probably Stones River age. Near the playa to the north of Goldbelt Spring Messrs. Chapman and Spaul-ding collected fossils of which the gasteropods, Mr. Ulrich states, remind one of Baltic Ordovician species and indicate a lower Ordo-vician horizon. The fossils include a very large *Helicotoma*-like shell; *Eccyliopterus* sp. undet., of large size and with contiguous whorls; *Receptaculites* sp. nov., obconical, with thick walls and small, crowded stolons. Not only in its organic remains, but also in its lithologic character, this limestone resembles the Pogonip limestone present in many of the ranges in the area surveyed. It was deposited in water of shallow or medium depth, a conclusion indicated by the presence here and there of cross-bedding and abundant organic re-mains. With the accumulation of the limy sediments the sea bottom must have been depressed step by step, an inference drawn from the great uniformity and thickness of the limestone.

Pennsylvanian limestone.—The Pennsylvanian limestone forms a narrow band along the east front of the Panamint Range. It is separated from the Pogonip limestone to the west by a fault of several thousand feet displacement. This limestone, of which probably some 1,500 feet is exposed, is fine grained and gray or dark gray in color. Some of the darker limestone when struck with a hammer gives off a fetid odor. Lenses and ellipsoid masses of black flint occur in some beds. To the west of the soda-syenite area is a bed 75 feet thick of white, fine-grained quartzose sandstone. This sandstone, as well as some beds of limestone, is cross-bedded.

From one locality the following Pennsylvanian fossils, determined by Dr. George H. Girty, were collected:

Fusulina sp.   Small indeterminable gasteropod.
Crinoidal fragments.   Indeterminable fragments.
Fenestella sp.

Fragmentary fossils were collected from two other localities, and concerning these Doctor Girty states that while they are undoubtedly Carboniferous, they indicate nothing further.

Older alluvium.—On either side of Emigrant Wash and at the north end of the Panamint Range are a number of areas of the older alluvium. The deposit east of Emigrant Wash forms intensely dissected hills which are continuations of the ridges of Tucki Mountain. The two large areas across the valley to the northwest are eastward-sloping plains cut by erosion into low hills, and each is separated from Death Valley by a low ridge of Paleozoic limestone. A few small outliers occur higher up in the mountains at a maximum altitude of 5,000 feet.

These deposits are composed of beds of angular bowlders and well-stratified sand. These are locally so well cemented by calcium carbonate that caves 25 feet high exist in them. Around Emigrant Spring, in the southern extension of the area west of Tucki Mountain and in the area at the north end of the Panamint Range, white and pink clay similar to that of the present playa deposits is interbedded with the bowlder beds. At the north end of the range the series has an exposed thickness of 2,500 feet.

Recent desert gravels unconformably overlie the older alluvium, which in turn lies upon the eroded surface of the quartz-monzonite and Paleozoic rocks. The upper portion of the series and the basalt are contemporaneous, and in consequence the older alluvium is probably of Pliocene and early Pleistocene age. In Pliocene-Pleistocene time, when Death Valley was occupied by an extensive playa, the Panamint Range was comparatively low and insignificant. The older alluvium areas are the remnants of alluvial slopes which once fringed its borders. These alluvial slopes near Goldbelt Spring reached an elevation of 5,000 feet, while to the north of Tin Mountain no remnants remain above 4,000 feet. Considerable portions of the Tin and Tucki mountain deposits are formed of playa clays, showing that during periods of exceptional rainfall the playa of Death Valley extended well up on the Panamint Range.

Recent alluvium.—The valley above the head of Cottonwood Creek is covered with coarse soil and loam. A more resistant portion of the quartz monzonite has narrowed and partially dammed the valley, and behind this barrier the material washed from the surrounding hills has accumulated. Four miles north of Goldbelt Spring is a basin whose bottom in depressed 275 feet below the inclosing rim of hills.

The basin is in Pogonip limestone and is probably the site of an ancient sink hole formed when the climate was less arid. Later the channel became obstructed by the inwash of débris from the surrounding hills, and eventually alluvial slopes and a playa were formed.

IGNEOUS ROCKS.

Earlier quartz-monzonite porphyry.—A few small fragments of a fine-grained, medium-gray igneous rock are included in the quartz monzonite in the southwest corner of the area mapped. Similar inclusions in the granite of Gold Mountain prove, on microscopic examination, to be quartz-monzonite porphyry.

Quartz monzonite.—A batholith of quartz monzonite underlies 75 square miles in the southwest corner of the area surveyed. The rock is light to medium gray in color, rarely pinkish gray. The minerals visible to the unaided eye are predominant gray feldspar, a greenish-black mineral, either amphibole or pyroxene, bronze-brown biotite, and pink feldspar, while small individuals of magnetite and seal-brown titanite are also visible in places. As a whole biotite is almost as abundant as hornblende, but it appears to be totally lacking over restricted masses. The diameter of the component minerals varies throughout the mass from one-fourth to one thirty-second of an inch. In many places the gray feldspars occur as well-developed tablets up to three-fourths of an inch in length and impart to the rock a porphyritic aspect. Green epidote in granules replaces the hornblende and biotite more or less completely, while veins of epidote cut the rock, and felts of this mineral are common on joint surfaces. Calcite veins are less common. The monzonite is cut by joint planes from 1 foot to 5 feet apart. In weathering the joint blocks become slightly rounded and finally disintegrate into soil. The monzonite forms rugged ridges. The outcrops are usually low bosses or bowlderlike masses, and many of the hills have the appearance of moraines, upon which large bowlders are prominent.

Under the microscope the rock shows an allotriomorphic granular texture, although some of the smaller plagioclase individuals form laths. The essential constituents, in order of abundance, are plagioclase, orthoclase, piotite, quartz, hornblende, and ugite. The accessory minerals are apatite, magnetite, and titanite. Disks of micropegmatite lie between the other constitutents and are inclosed in the feldspars. Some hornblende is secondary to augite, while kaolin, sericite, and epidote form at the expense of the feldspars. The rock is on the border line between quartz monzonite and granodiorite.

The monzonite is cut by thin dikes of a fine-grained pink aplite composed of feldspar and some hornblende. Quartz is sometimes seen in the rock, which then appears to be of granitic composition. The dikes are usually simple. Under the microscope this rock shows the composition of an acidic quartz monzonite, containing a little biotite altered to chlorite. The accessory minerals are the same as those of the quartz monzonite. Some of the aplite is blotched by globular aggregates of coarser hornblende from 1 inch to 4 inches in diameter. These aggregates are formed by the segregation of hornblende from the surrounding rock and are rimmed by halos of almost pure feldspar in individuals from one-half to 1 inch wide. The aplite is probably but a later differentiation product of the monzonitic magma.

Scattered throughout the monzonite are segregations of hornblende from 2 to 3 inches in diameter. These represent a threefold enrichment in hornblende by magmatic differentiation.

The monzonite is cut by and grades into thin dikes of coarser grained rock, locally having the same composition as the monzonite, but usually of more acidic constitution. The individuals of these pegmatite dikes reach a maximum diameter of 2 inches. Some of the dikes have basic borders composed of almost pure hornblende, with a median band of feldspar. While the boundary planes are rather distinct, individual crystals extend from the monzonite into the basic bands and from these into the acidic center. The more acidic portion of this pegmatite shows under the microscope the composition of a soda syenite, containing orthoclase, anorthoclase, augite, hornblende, biotite, quartz, titanite, magnetite, apatite, and zircon. The augite probably contains a little of the ægirite molecule. Other forms of the pegmatite are composed of quartz, pinkish-gray feldspar, and hornblende, named in the order of abundance and the reverse order of solidification. Some limonite pseudomorphs after pyrite occur in the pegmatite, but the original character of the pyrite was not determined beyond doubt. In other pegmatites partial crystals of ferromagnesian minerals are inclosed in a coarse crystalline aggregate of feldspar. Microscopic examination showed that orthoclase and microperthite are so predominant over plagioclase in this rock that it has the composition and texture of a syenite porphyry. The ferromagnesian minerals include augite, olivine, brown hornblende, and biotite. Apatite occurs in unusually large crystals and magnetite in grains. Miarolitic openings occur in this pegmatite and crystals of the constituent minerals protrude into them. The pegmatite therefore consolidated under less pressure than the monzonite, which lacks such cavities, or its composition was more favorable to their production. The pegmatite is evidently but a late intrusion of the monzonite-magma residuum. Not only is the pegmatite more acidic than the quartz monzonite, but it has undergone an enrichment in soda, linking the quartz monzonite to the soda syenite next to be described.

The batholith sends thin dikes into the Pogonip limestone. At the contact the limestone is considerably disturbed and dips sharply, usually away from the monzonite mass. The small limestone area inclosed by monzonite on the southern border of the area appears to have been buoyed up on the surface of the molten rock. Smaller inclusions are common. An aureole of limestone from one-fourth to 1 mile wide is metamorphosed to a rather coarse-grained gray marble. In this marble are bands, lenses, and irregular bodies of white marble. Large brown garnets, epidote, serpentine, and tremolite are present in the metamorphosed limestone. The two latter minerals occur in veins as well as in sporadic masses through the limestone.

The quartz monzonite intrudes Pogonip limestone and is cut by pre-Tertiary diorite porphyry. It is presumably contemporaneous with the post-Jurassic granite, since intermediate facies between it and the granite, described on page 208, appear to exist. It is rather similar in mineralogic composition to the post-Jurassic granodiorites of the Sierra Nevada.

Soda syenite.—About $11\frac{1}{2}$ miles southeast of Tin Mountain there is a very irregular mass of soda syenite and soda-syenite porphyry, 300 feet long and 200 feet wide. It intrudes the Pennsylvanian limestone and sends into it many straight-walled dikes and ramifying veins of varying width. Prior to recent erosion the limestone appears to have covered the syenite. Bowlders of similar rock occur in the gulch due west of this area, but the masses from which these were derived were not located. Similar rock appears from a distance to form a small area 4 miles west of Lost Wagons. The soda syenite is characterized by abrupt and great changes in granularity and in the relative abundance of the constituent minerals. The predominant form is a coarse- to medium-grained rock of gray color, composed of predominant gray with some pink feldspar, subordinate greenish-black amphibole or pyroxene, and black mica. Many of the feldspars have good crystal outlines and in the more porphyritic facies the abundant feldspar laths have a length of $1\frac{1}{2}$ inches and are aligned in flow orientation. The rock next to the limestone is very fine grained. Epidote has developed at the expense of the hornblende and biotite in all facies. Under the miscroscope this rock proves to be a soda syenite or nordmarkite of hypidiomorphic and uneven granular texture. The predominant constituents are alkali feldspars, including the species orthoclase, microperthite, and anorthoclase. With these is a little oligoclase. The alkali feldspars form rude tabular crystals, many of which are twinned according to the Carlsbad law. Between these tabular forms are anhedra of augite, quartz, biotite, and yellowish-brown garnet. The augite verges toward ægirite-augite. The accessory minerals are titanite, magnetite, apatite, and fluorite. The fluorite flecks the alkali feldspar and may be original or introduced by magmatic gases.

The soda syenite is cut by dikes of compact greenish-gray aplitic rocks. Pink feldspar is associated with the gray feldspar and pyroxene and it alone forms some veins. The aplite under the microscope has a very uneven-grained allotriomorphic texture. The feldspars include microperthite, orthoclase, and some anorthoclase. A little quartz is also present. While the rock is not rich in ferromagnesian minerals, there are many small grains and partial crystals of ægirite-augite. This mineral shows the usual zonal structure, with deeper green bands on the border. Irregular grains of a light-yellow garnet are rather abundant. The accessories are titanite and apatite. Fluorite occurs in small blebs, surrounded by a mesh of sericite shreds, which were probably formed by the same gases that deposited the fluorite. Calcite is also present and its contacts with other minerals are so sharp that it was probably deposited in miarolitic openings. The contact between the aplite and the syenite is in some cases sharp, in others gradational. In one instance a dike of syenite porphyry is faulted by an aplitic dike, but there can be little doubt that the two are genetically related. Narrow pegmatitic dikes, the feldspar and pyroxene of which reach a diameter of 1 inch, occur.

The limestone immediately adjoining the igneous mass is metamorphosed to a coarse- or fine-grained white marble, through which small plates and veinlets of tremolite are scattered.

The soda syenite intrudes Pennsylvanian limestone and has suffered practically the same deformation as the quartz monzonite, and it may well be, as indicated by the monzonite-pegmatite, a later variant of the same magna.

Granite.—A considerable area of biotite granite occurs to the east of the Emigrant Wash, 3 miles south of the area surveyed. Dikes of muscovite granite, genetically related to the biotite granite, were observed cutting the Cambrian rocks 1 mile south of Tucki Mountain, and a few probably occur in the area surveyed. This granite is probably of post-Jurassic age.

Later quartz-monzonite porphyry.—Four miles south of Tin Mountain is an area of siliceous quartz-monzonite porphyry, probably intrusive in the Pogonip limestone. This area was not seen by the writer, but specimens collected by Mr. R. H. Chapman closely resemble the white quartz-monzonite porphyry of the Silver Peak Range, and, like it, this rock is probably genetically related to the intrusion of the post-Jurassic granitoid rocks. Under the microscope the phenocrysts of quartz, orthoclase, plagioclase, and biotite are seen to be embedded in the finely granular groundmass of microcline, plagioclase, quartz, and biotite. The accessory minerals are magnetite, titanite, and apatite.

Diorite porphyry.—Dikes of diorite porphyry from 2 to 6 feet wide cut the quartz-monzonite batholith in the southwest corner of the area surveyed. The rock is light greenish gray in color and has prominent white feldspar phenocrysts from one-eighth to one-fourth inch in length, which are in parallel alignment through flow. The rock has suffered considerable epidotization. Under the microscope it proves to be a much-altered diorite porphyry with a groundmass of plagioclase laths and a little orthoclase. Aggregates of epidote, zoisite, and chlorite and a smaller amount of calcite are evidently pseudomorphs of plagioclase and hornblende phenocrysts. This rock is evidently the pre-Tertiary diorite porphyry.

Basalt.—Six small masses of basalt occur in the southern portion of the Panamint Range, and the same rock caps a considerable area at the north end of the range, where it occurs in several flows, each 500 or more feet thick, interbedded with the older alluvium. The small basalt mass 1¾ miles south of west of the head of Cottonwood Creek is a north-south dike 250 feet wide. The basalt on the borders of this dike is vesicular and includes numerous quartz-monzonite fragments. The outcrop on the Cottonwood Canyon trail three-fourths of a mile above its entrance into the mountains may also be a dike cutting the Carboniferous limestone. The other masses are erosional remnants of a single flow or of contemporaneous flows. The prebasaltic mature Panamint Range in consequence must have been cut deeply by canyons on its borders, while the old surface to the south and southwest of the area mapped was but little dissected when the basalt was extruded. The flow on the east side of Emigrant Wash forms a black bluff 200 feet high, which is a prominent landmark for many miles. The flow 4 miles east of the dike is 400 feet thick and forms a mesa.

The basalt has a dense dark-gray groundmass mottled by light-gray areas. It is more or less vesicular, the largest cavities being 2 inches long and usually elongated parallel to the direction of flow. Calcite, quartz, and zeolites fill the vesicles more or less completely. The phenocrysts, which are usually subordinate to the groundmass, include blebs of olivine, more or less altered, and striated feldspars up to one-fourth inch in length, some of them showing zonal growth. Greenish-black augite is less commonly present. Under the microscope the basalt shows a dark, greasy groundmass, in which are microlitic laths of basic plagioclase and pyroxene columns. The phenocrysts include laths of basic plagioclase; rounded grains of olivine, usually fresh, although in places altered to iron-stained serpentine, and columns of grayish pyroxene, much of it twinned. The basalt is usually well jointed and weathers into spheroidal masses.

The basalt east of Emigrant Wash and the flow on the Cottonwood Canyon trail lie unconformably below the older alluvium. The

flows at the north end of the range and others near Emigrant Spring
are interbedded with the same formation. The deposition of the
upper part of the older alluvium and the effusion of the basalt were
therefore contemporaneous, although the latter process continued
after the former had ceased. The basalt is probably of the late Pliocene and early Pleistocene age.

STRUCTURE.

The structural history of the Panamint Range has doubtless been
as complex as that of the Amargosa Range, but the absence of
Miocene formations removes the means by which the pre-Tertiary
and middle Tertiary folding can be differentiated. It is, however,
evident that the Paleozoic rocks were folded prior to the intrusion
of the post-Jurassic igneous rocks.

Tucki Mountain appears to be the northward-pitching end of a
north-south canoe, the western side of which is buried beneath
Emigrant Wash or has been faulted off. The Cambrian rocks show
well the differential folding suffered by a heterogeneous rock series.
The schist is typically closely folded and even crenulated; the
limestone bands are less strongly folded while the folds of the
quartzite are open. The lithologic character of the series renders
the detection of the many faults difficult. The folding of the Pogonip and Pennsylvanian limestones is complex, although in its minor
details less intense than that of the Amargosa Range, since overturned folds and isoclinals are unusual. The major fold is an anticline, the axis of which is situated near the western border of the
area and trends a little west of north. The eastern arm of the fold
is long and dips gently eastward. Superimposed upon it are many
minor parallel folds, while cross folds with east-west axes cross it.
This fold appears to die out a short distance south of Tin Mountain
and here the strata are approximately horizontal. The earlier folding was apparently accompanied by reverse strike faulting. The
intrusion of the quartz monzonite in the southwest corner of the
area mapped considerably disturbed the limestones in its immediate
vicinity. They are buckled and dip steeply, usually away from the
batholith. Faulting has occured at the contact, as well as in the
quartz-monzonite mass itself.

The fault between the Pogonip and Pennsylvanian limestones is
presumably a normal fault and is certainly of several thousand feet
displacement. Its position is only approximately shown on the
map. Small normal faults of northwesterly strike are common in
the Pennsylvanian limestone near Death Valley, and these were
probably formed when the valley originated. Similar faults with
2 to 4 feet displacement occur in the older alluvium, and this formation has been uplifted and tilted toward Death Valley in places as
much as 15°.

No mining camp exists in the portion of the Panamint Range included in the area under consideration. Harrisburg lies near the Emigrant Spring-Ballarat road, about 6 miles south of the boundary. Assays reported by prospectors indicate that gold-bearing copper ores occur in the quartz monzonite and silver-bearing galena ores in the limestone areas.

Goldbelt, a deserted camp, is situated at the Goldbelt Spring, near the contact of the Pogonip limestone, here marmorized, and the quartz monzonite. A number of men are said to have rushed to this camp in the spring of 1905, but little work was done. The more important development work was on certain thin veins or lenses in the quartz monzonite. The ore contains a little chalcopyrite, probably a portion of the original sulphide unaltered. Of later origin are malachite, chrysocolla, and a dark-brown or black, iron-stained, flinty chalcedony, of approximately contemporaneous age. These minerals are coated with small quartz crystals. This ore is said to have panned free gold. Similar ore is reported from other localities in this monzonite area. A specimen from a vein near the source of Cottonwood Creek consists of dark limonite-stained jasperoid and chrysocolla, in which occur radial crystals of bottle-green brochantite. This also is reported to pan gold. In the soda-syenite mass $6\frac{1}{2}$ miles southeast of Tin Mountain thin veins of limonite, presumably after iron sulphide, occur.

The areas most favorable to prospecting for precious metals are the Cambrian rocks of Tucki Mountain and the Pogonip limestone in the vicinity of the quartz-monzonite mass. Quartz veins, usually of lenslike form, are common on Tucki Mountain. They reach an observed maximum width of 3 feet and usually appear barren, although in some instances they have been crushed and are heavily stained by limonite and hematite. It is said that an old lead mine was at one time worked on this mountain, and several promising camps in the Amargosa Range are situated on similar veins in the same rocks. Veins of two ages in the Pogonip limestone are formed of barren-looking calcite. Some occupy overthrust fault fissures, and these are faulted by other veins. The older veins are composed of massive crystalline calcite; the younger are often beautifully crustified. On the Cottonwood Canyon trail magnificent calcite veins 4 feet wide occupy fault fissures. The crustified calcite curves around included fragments of limestone in concentric bands. Near the quartz-monzonite masses a few small quartz veins occur and some of these contain pyrite. In other portions of the area surveyed ore-bearing veins have been found in limestone near granite contacts, and by analogy they probably also exist near the related monzonite. At a number of places in the limestone the rock is heavily stained by hematite and limonite. In some cases at least these appear to be gossan deposits and the determination of the underlying sulphides might prove profitable.

The older alluvium is contemporaneous with rocks of Death Valley in which some borax minerals have been found. With the possible exception, however, of the area north of Tin Mountain and one up Emigrant Wash beyond the boundary, the Panamint Range is not worth prospecting for borax.

U. S. GEOLOGICAL SURVEY   BULLETIN NO. 308 PL. III

MARBLE CANYON, PANAMINT RANGE.

Showing metamorphosed Pogonip limestone near quartz-monzonite batholith.

INDEX.

www.ingramcontent.com/pod-product-compliance
Lightning Source LLC
Chambersburg PA
CBHW070515200326
41519CB00013B/2809